Intangibles in the Big Picture: The Delinearised History of Time

INTANGIBLES IN THE BIG PICTURE: THE DELINEARISED HISTORY OF TIME

GARY ZATZMAN
AND
RAFIQUL ISLAM

Nova Science Publishers, Inc.
New York

Copyright © 2009 by Nova Science Publishers, Inc.

All rights reserved. No part of this book may be reproduced, stored in a retrieval system or transmitted in any form or by any means: electronic, electrostatic, magnetic, tape, mechanical photocopying, recording or otherwise without the written permission of the Publisher.

For permission to use material from this book please contact us:
Telephone 631-231-7269; Fax 631-231-8175
Web Site: http://www.novapublishers.com

NOTICE TO THE READER

The Publisher has taken reasonable care in the preparation of this book, but makes no expressed or implied warranty of any kind and assumes no responsibility for any errors or omissions. No liability is assumed for incidental or consequential damages in connection with or arising out of information contained in this book. The Publisher shall not be liable for any special, consequential, or exemplary damages resulting, in whole or in part, from the readers' use of, or reliance upon, this material. Any parts of this book based on government reports are so indicated and copyright is claimed for those parts to the extent applicable to compilations of such works.

Independent verification should be sought for any data, advice or recommendations contained in this book. In addition, no responsibility is assumed by the publisher for any injury and/or damage to persons or property arising from any methods, products, instructions, ideas or otherwise contained in this publication.

This publication is designed to provide accurate and authoritative information with regard to the subject matter covered herein. It is sold with the clear understanding that the Publisher is not engaged in rendering legal or any other professional services. If legal or any other expert assistance is required, the services of a competent person should be sought. FROM A DECLARATION OF PARTICIPANTS JOINTLY ADOPTED BY A COMMITTEE OF THE AMERICAN BAR ASSOCIATION AND A COMMITTEE OF PUBLISHERS.

LIBRARY OF CONGRESS CATALOGING-IN-PUBLICATION DATA

ISBN 978-1-60692-249-1

Available upon request

Published by Nova Science Publishers, Inc. ✣ New York

Contents

Chapter 1	Introduction	1
Chapter 2	Newton's "Laws of Motion" – versus Nature's	3
Chapter 3	From Illusions of Precision and Reproducibility in Natural Science to Delusions of Normalcy in Social Science	13
Chapter 4	Mutability	23
References and Bibliography		85
Index		121

Chapter 1

INTRODUCTION

The so-called "TINA syndrome" provides the fundament, the actual rock, on which the political, economic, military and other elites and establishments of the Anglo-American world and European bloc have built their church. Inscribed over its entrance stands the motto: "there is no god but monopoly and maximum is his profit". On this basis, continuous attacks on the very concept of intangibles are launched, most prominently against time-consciousness. Especially singled out is time-consciousness based on appreciating and-or priorising the long term over the short term, as well as placing the interests of the social collective over the interests of any individual member of the collective. In this portion of the chapter, it is argued that Humanity has been on the wrong track since Sir Isaac Newton published his Principia Mathematica at the end of the 17th century, and that the scientific research enterprise developed since then has taken the world on a merry chase to nowhere.

Without exception, the assaults on time-consciousness, and on cognition of what happens in and through the passage of time, take the form of a denial of the principle of Nature as the Mother of all wealth. The denial of this principle has always encountered resistance. Some resist by breaking the attacks down and responding to selected cases. For example, the contributors to the book Underdevelopment and Social Movements in Atlantic Canada (Toronto 1979), following precisely this tact, act according to the principle that "the movement is everything…" This places the struggle of the people for livelihood where it belongs, viz., at the centre of economic theory and practice. However, these writers' version of this approach is silent about long-term or final aims. Their work actually priorises t = "right now" over longer-term views of the role of time in social-historical processes. People's deepest desires to see Justice prevail and Injustice sent packing are generally aroused, positively, by their apparent stand on the side of "labour" against "capital"; a great deal of hope might well be vested in these stands. Has this hope, however, been misplaced? Analysis of these authors' collective work from 1979 (as the Soviet Union began its final slide to oblivion by invading Afghanistan), and its source in theories of "regional underdevelopment" (formulated at the Cold War's height in the late 1950s), suggests this may be the case. Especially disturbing is the outlook underlying that theory, and specifically its extreme pragmatism and welter of contradictions and inconsistencies. These disclose a position entirely at odds with the proclaimed mission to establish the truth of matters under investigation.

In order to maintain a position in what they see as the mainstream today, some of these writers have taken matters further, adapting to fit the cut of current discourse in the early 2000s some of the concerns raised in the earlier work. En route, however, they make a major concession to the disinformation of the Canadian fisheries department that "there are too many fishermen chasing too few fish". Disguising the concession as an appeal for "ecological sanity" in the face of a pending environmental crisis of raw material food supplies during a period of still-excessive capitalization in the coastal fishing industry, those putting forward this argument decline to challenge the claims by the government and the largest fish processors that the problem at bottom is a shortage of raw material, a defect in Nature. As, however, the problem is actually one of how Humanity has arranged its affairs when it comes to extremely fundamental matters like food-gathering, this concession, no less than any of the other more direct attacks on time consciousness and on cognition as a source of reliable information, forms part of a far more general and sweeping assault on the very concept of human agency. This assault challenges the fundamental notion that no human social problem is without some human social solution. The fact of the matter is that the essence of human social agency lies on the path of pursuing knowledge. Whosoever would increase knowledge is bound to disturb the status quo, but even so, a person must increase his knowledge of the truth.

Chapter 2

NEWTON'S "LAWS OF MOTION" – *VERSUS* NATURE'S

As an enterprise entailing apprehension and comprehension of the material world existing external to consciousness, science – meaning the scientific approach to investigating phenomena – requires examining both things-in-themselves and things in their relations to other things.

One of the most fundamental nuts to be cracked in this exercise involves mastering and understanding laws of motion as they apply to the matter under investigation. The importance is simply that motion is the mode of existence all matter. Whether it is energy, or matter that has become transformed into energy, or energy that became transformed into matter, there is no form of material existence that is not in motion.

There are two variables that have become especially critical for modelling and tackling the actual laws of motion of modern economic life: time and information. Both are utterly intangible. Up to now, however, they have been incorporated into economic analysis on the basis of rendering them tangible. This has created more problems than it solved.

2.1. THE CONTINUITY CONUNDRUM

On the front of scientific work undertaken to investigate and determine laws of motion, Isaac Newton represents the watershed. His elaboration of the general laws of motion of all matter was a huge advance over the incoherent and conflicting notions that prevailed hitherto. Of course, various limitations appeared at certain physically measurable/detectable boundaries – at speeds approaching the speed of light, for example, or within space approaching the measurable minimum limit of (approximately) 10^{-32} m, etc. This led researchers to make important corrections and amendments to Newton's formulae. The fact remains, nevertheless, that Newton's fundamental breakthrough lay in the very idea of summarising the laws of motion itself, common to all discrete forms of matter understood and observed to that time, *i.e.*, not atomic, molecular or sub-atomic. Equally remarkably, in order to take account of the temporal component attending all matter in motion, Newton invented an entirely new departure in mathematics. A new departure was required because existing mathematics were useless for describing any aspect of change of place while matter was undergoing such change.

Apart from their long standing despite some amendment, this mathematical apparatus used to describe and apply Newton's laws is worth re-examining to get a better understanding of some of the basic tools used throughout scientific work in all fields, including fields far removed from having to deal with laws of motion. Here we have in mind the fundamentals of integral and differential calculus.

Newton's mathematics made it possible to treat time as though it were as infinitely divisible as space – something no one had ever conceived of doing before. This worked extremely well for purposes involving the relative motion of masses acting under the influence of the same external forces, especially force due to gravity and acceleration due to gravity. Extended to the discussion of the planets and other celestial bodies, it appeared that Time throughout nature – Time with a capital "T" – was indeed highly linear. For Newton and for all those applying the tools of his calculus to problems of time and space comprehensible to ordinary human perception, t_{LINEAR} and $t_{NATURAL}$ were one and the same.

Newton's was an extremely bold and utterly unprecedented maneuver. It arrived as the fruit of an unpredicted turn in the profound revolution in human thought already under way since the start of the Renaissance during the century and a half predating Newton. Launched from the leading centres of the Bourbon and Hapsburg Empires, to reverse the correct verdicts of the new science of Copernicus, Kepler, Galileo and others that emerged during the European Renaissance in increasingly open revolt against the authority of Church dogma, the Catholic counter-reformation had failed, and failed utterly. Throughout the continent of Europe, Catholic monarchs and the authority of the Holy Roman Catholic Church were placed entirely on the defensive. In England, the "Catholic forces" were entirely routed, and among that country's scientific and philosophical circles, Newton, along with many of his colleagues in the Royal Society, were standard-bearers of the newly-victorious forces.

Newton's mathematical labour was nothing like the mystical, quasi-religious revelation that his fellow eighteenth-century Englishman, Alexander Pope, captured in the line "God said: 'Let Newton be', and all was light." To elaborate his method into what he called, in the *Principia Mathematica*, a "theory of fluents and fluxions", Newton built on and refined the implications and tentative conclusions of a number of contemporaries and near-contemporaries who, although lacking an overarching theoretical framework, were already working with processes of infinite summation that converged to some finite value. He proposed differentiation as a method for deriving rates of change at any instant within a process, but his famous definition of the derivative as the limit of a difference quotient involving changes in space or in time as small as anyone might like, but not zero, *viz.*:

$$\frac{d}{dt}f(t) = \lim_{\Delta t \to 0} \frac{f(t+\Delta t)-f(t)}{\Delta t}$$

Figure 2-1. Formulation of Newton's breakthrough idea (expressing Leibniz' derivative notation in Cauchy's "limits" notation)

set the cat among the pigeons. It became apparent that, without further conditions being defined as to when and where differentiation would produce a meaningful result, it was entirely possible to arrive at "derivatives" that would generate values in the range of a function at points of the domain where the function was not defined or did not exist. Indeed: it took another century following Newton's death before mathematicians would work out the conditions – especially the requirements for continuity of the function to be differentiated within the domain of values – in which its derivative (the name given to the ratio-quotient generated by the limit formula) could be applied and yield reliable results.

Only seven years after Newton's death, the main arguments hoisted against his mathematics and especially some of its underlying, implicit notions came not from scientists but from Christian theologians, led by Church of England bishop George Berkeley. He was the most prestigious among those who considered inherently blasphemous the very idea that mental apparatus of the human could aspire to manipulate and control any infinite process. Although such an idea would be unlikely to occur to modern reader, not all the Bishop's remarks were without merit. Deriding Newton's differentials as "ghosts of vanishing quantities", Berkeley (1734) encapsulated the actual problem in terms that would echo among mathematicians for another century. Others on the same line all but accused Newton of poaching on The Infinite as a supposedly exclusive turf of the

Almighty. Newton's public posture was that men could know a Divine plan for Nature through grasping the physical laws, but he declined to publish his own views more fully. From his private papers it is now known that he saw knowledge of these laws mainly as a potential source for individuals to enrich themselves. His motive in uncovering natural laws was also in part linked to the desire he shared with many other English men of science of his day to discredit those doctrines – especially concerning the nature of matter, motion, and planetary bodies – whose sole support rested on the authority of the Catholic Church and its papal index.

2.2. CONTINUITY AND LINEARITY: CONFUSION TWICE CONFOUNDED

It was in the period 1740-1820 that the basic theory of differential equations also came to be elaborated. Newton's notation was almost universally replaced by that of calculus' cofounder Leibniz, facilitatiing the achievement of several further breakthroughs in the theory of analysis for the Swiss mathematician Euler among others. Many notable techniques were developed using the techniques of superposition (Kline 1972).

The notion of superposition was an ingenious solution to a very uncomfortable problem implicit in (and left over from) Newton's original schema. Under certain limiting conditions, his derivative would be useful for dealing with what today are called vectors – entities requiring at least two numerical quantities to fully describe them. All the important and fundamental real-world entities of motion – velocity, acceleration, momentum etc – are vectorial insofar as, if they are to usefully manipulated mathematically, not only their magnitude but also their direction must be specified.

Here there inheres an limiting condition for applying Newton's calculus. So long as magnitude and direction change independently of one another, no problems arise in having separate derivatives for each component of the vector or in superimposing their effects separately and regardless of order. That is what mathematicians mean when they describe or discuss Newton's derivative being used as a "linear operator". The moment it is not possible to say whether these elements are changing independently, however, a linear operation will no longer hold. Because modelling is always an approximation, this for a long time provided many researchers a licence to simplify and relax requirements, to some degree or other, as to just how precisely some part of natural reality had to fit the chosen or suggested model. Naturally, one could generate some sort of model, and results, provided the assumptions – boundary conditions or initial conditions – were then retrofitted more or less so as to exclude unwanted dependencies. The interior psychology of this act of choice seems to have been that the linearised option would reach a result, therefore it could and should be used. The implication of this choice has been rather more mischievous: everything non-linear has been marginalised either as exceptional, excessively intractable in its "native" non-linear state, or usable only insofar as it may be linearised.

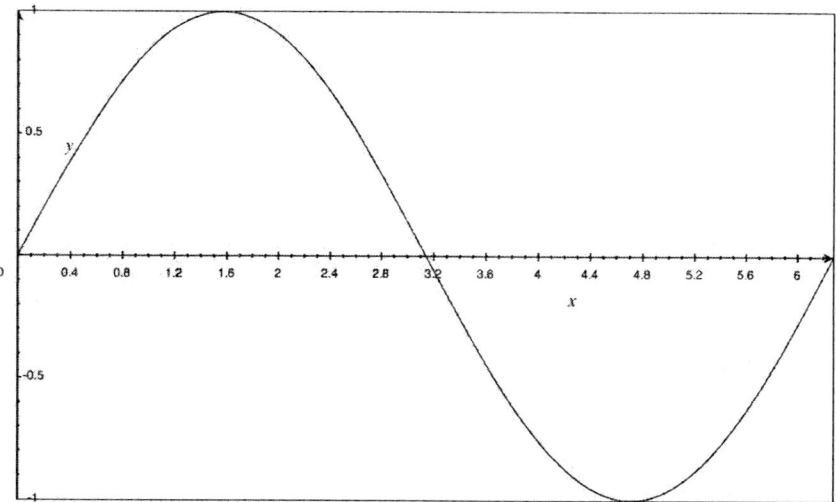

Figure 2-2. Graphic representation, in Cartesian coordinates, of the classic simple function $f(t)=sin\ t$.

In the actual evolution and development of what became the field of real analysis, of course, every step was taken incrementally. Newton's discoveries were taken up and re-used as tools. Meanwhile, however, the theoretical work needed to explain the conditions under which analytic methods in general, and the derivative in particular, were applicable had not reached the stage of explicit elaboration. Thus, the notion of the derivative as a linear operator, and even aspects of a more generalised theory of linear operators, began to develop and be utilised before the continuity criteria underpinning the entire field of real analysis were made explicit. This led to associating linearity principally with superposition techniques and the possibility of superposition. By the time Cauchy published his work elaborating the importance of continuity, no one would connect continuity with linearisation. In real analysis, as Kline (1972) and most other modern historians of mathematics have observed, discontinuity became correlated mainly and even exclusively with undifferentiability.

With the rigourising of real analysis by Cauchy and Gauss, applied mathematics in the middle third of the nineteenth century developed a powerful impetus and greatly broadened its field of action throughout all the natural sciences, especially deeply in all areas of mechanical engineering. There arose a preponderating interest in steady and-or equilibrium states, as well as in the interrelations between static and dynamic states.

While this was not at all unexpected, it is crucial at this point to make what was actually going on more explicit. Some initial analysis of a deliberately simplified example (see Figure 2-2) will help illuminate something that often becomes obscured:

Assume some process described by the simple sine function illustrated above. As may be recalled from introductory calculus, using Newton's difference-quotient formula (from Figure 1), the instantaneous rate of change anywhere along the graph-line of this function, which will be continuous anywhere within the interval $(-\infty, +\infty)$, i.e., $-\infty \le t \le +\infty$, can be computed stepwise as follows:

$$\frac{d}{dt}f(t) = \lim_{\Delta t \to 0} \frac{\sin(t+\Delta t) - \sin t}{\Delta t} = \lim_{\Delta t \to 0} \frac{\sin t \cos \Delta t + \sin \Delta t \cos t - \sin t}{\Delta t}$$

As Δt approaches 0, $\cos \Delta t$ approaches $\cos 0$, which is 1. Meanwhile, because $\sin x$ approaches x for decreasingly small values of x, the term $\frac{\sin \Delta t}{\Delta t}$, also becomes unity. So:

$$\frac{d}{dt}f(t) = \lim_{\Delta t \to 0} \frac{\sin(t+\Delta t) - \sin t}{\Delta t} = \lim_{\Delta t \to 0} \frac{\sin t \cos \Delta t + \sin \Delta t \cos t - \sin t}{\Delta t} =$$

$$\lim_{\Delta t \to 0} \frac{\sin t + \sin \Delta t \cos t - \sin t}{\Delta t} =$$

$$\lim_{\Delta t \to 0} \frac{\sin \Delta t \cos t}{\Delta t} = \lim_{\Delta t \to 0} \frac{\sin \Delta t}{\Delta t} \cos t = \lim_{\Delta t \to 0} \cos t = \cos t$$

$$\lim_{\Delta t \to 0} \frac{\sin \Delta t \cos t}{\Delta t} = \lim_{\Delta t \to 0} \frac{\sin \Delta t}{\Delta t} \cos t = \lim_{\Delta t \to 0} \cos t = \cos t$$

Figure 2-3. Generating the first derivative $f'(t)$ for the function $f(t) = \sin t$ using Newton's difference-quotient formula.

This means that, as one moves continuously along the domain t, the *instantaneous rate of change* along the curve represented by the graph for f(t) can be computed by evaluating the cosine of t at that value on the horizontal axis. What is being described is change *within* the function; *the function itself, of course, has not changed.* As this particular function happens to be periodic, it will cycle through the same values as the operational output described by this graph proceeds through subsequent cycles. This makes it quite easy to see that the function itself describes a steady-state condition. In fact, however, even if the function were some polynomial, anything lying on the path of its graph would represent the steady-state operation of that function: steadiness of state is not reducible to some trait peculiar to periodic functions.

Newton's method itself, long described as "Newton's method of tangents" because it could be illustrated geometrically by picturing the derivative as the slope of a straight-line segment tangent to the curve of any function's graph, relies implicitly on the notion of approximating instantaneous moments of curvature, or infinitely small segments, by means of straight lines. This alone should have tipped everyone off that his derivative is a linear operator precisely because, and to the extent that, it examines change over time (or distance) *within an already established function, i.e.,* within a process that has reached its steady state.

The drive to linearise covers a multitude of sins. Thus for example, as bold and utterly unprecedented Newton's approach, it also contains a trap for the unwary: going backward or forward in space or in time is a matter of indifference. If natural reality is to be modelled as it actually unfolds, however, the requisite mathematics has to close the door on, and not permit the possibility of, treating time as reversible. What use can be made, then, of such mathematics for describing anything happening in nature according to naturally-conditioned temporal factors? To engineer anything in Nature, applying Newton's calculus requires suppressing or otherwise sidelining such considerations, and indeed: it has long been accepted, as a pragmatic matter, that fudge factors and ingenious work-arounds are needed to linearise the non-linear. What has not been clarified, or discussed much if at all up to now, is that this is *inherently* what they must be about. If this nub of the issue is inherent, then it follows that merely backing up a few steps on the path that brought matters to this stage, back to the point where everything still looked more or less linear and the non-linearities had not yet taken over, is not going to overcome the fundamental difficulty. The starting-point itself contains the core of the problem, which is that Newton's calculus edifice, in its very foundation, is truly anti-Nature. Starting anywhere on this path, one will diverge ever further from Nature.

One starting point for a new path might go somewhat as follows: consider as the starting point for modeling such natural processes some series of observations of an ongoing phenomenon for which there is no "analytic" function that fits perfectly or even fits over an extended run of results. Results are needed that are grouped reasonably closely in time (the assumption of continuity must be more-or-less likely or possible to validate). Instead, however, of computing a difference quotient based on evaluating the limit at some arbitrary common value like 0, consider what happens if some positive finite constant real value c were used instead. A new derivative may be defined thus:

$$f'(t) = \frac{d_c}{dt} f(t) = \lim_{\Delta t \to c} \frac{\sin(t+\Delta t) - \sin t}{\Delta t} = \lim_{\Delta t \to c} \frac{\sin t \cos \Delta t + \sin \Delta t \cos t - \sin t}{\Delta t}$$

Now, as Δt approaches c, $\cos \Delta t$ approaches $\cos c$, which is anywhere in the interval $[-1,+1]$. Meanwhile, the term $\frac{\sin \Delta t}{\Delta t}$ may fall anywhere in the interval $[-\frac{\sqrt{3}}{2}, +\frac{1}{c}]$. Applying these maxima and minima generates the open interval $-(2 \sin t + \frac{1}{c} \cos t) \leq \frac{d_c}{dt} f(t) \leq \frac{1}{c} \cos t$, in which:

- at $t = 0$ $(+2k\pi)$, $\frac{d_c}{dt} f(t)$ converges to a single value, viz., $\frac{1}{c}$, which is positive (> 0);

- at $t = \frac{\pi}{6}$: $-(2 + \frac{\sqrt{3}}{2c}) \leq \frac{d_c}{dt} f(t) \leq \frac{\sqrt{3}}{2c}$, which straddles 0;

- at $t = \frac{\pi}{4}$: $-\sqrt{2}(1+\frac{1}{2c}) \leq \frac{d_c}{dt} f(t) \leq \frac{\sqrt{2}}{2c}$, which straddles 0;

- at $t = \frac{\pi}{3}$: $-(\sqrt{3}+\frac{1}{2c}) \leq \frac{d_c}{dt} f(t) \leq \frac{1}{2c}$, which straddles 0;

- at $t = \frac{\pi}{2}$: $-2) \leq \frac{d_c}{dt} f(t) \leq 0$, which is mainly negative (≤ 0);

- at $t = \frac{2\pi}{3}$: $-(\sqrt{3} - \frac{1}{c}) \leq \frac{d_c}{dt} f(t) \leq -\frac{1}{2c}$, which is entirely negative (< 0);

- at $t = \frac{3\pi}{4}$: $-\sqrt{2}(1-\frac{1}{2c}) \leq \frac{d_c}{dt}f(t) \leq -\frac{\sqrt{2}}{2c}$, which is entirely negative (<0)

and reduces,

- at $c=1$, to $-\frac{\sqrt{2}}{2}$.

From here, heading towards $t = \pi$, other features emerge:

- At $t = \frac{5\pi}{6}$, $\frac{d_c}{dt}f(t)$ lies somewhere between $-(1 - \frac{\sqrt{3}}{2c})$ and $-\frac{\sqrt{3}}{2c}$, in

which:

- for $c = \frac{\sqrt{3}}{2}$, $-1 \leq \underset{\Delta t \to c}{f'(t)} \leq 0$;
- for $\frac{\sqrt{3}}{2} < c < \sqrt{3}$, while for $c > \sqrt{3}$, $\underset{\Delta t \to c}{f'(t)} < 0$; and
- for $c = \sqrt{3}$, $\underset{\Delta t \to c}{f'(t)} = -0.5$;
- At $t = \pi$: $-\frac{1}{c} \leq \frac{d_c}{dt}f(t) \leq \frac{1}{c}$

Figure 2-4. Generating family of first derivatives, $\{\underset{\Delta t \to c}{f'(t)} = \frac{d_c}{dt}f(t)\}$, for $f(t) = \sin t$ using modified difference-quotient.

Here we are dealing with multiple, in fact: infinite, solutions, as should be expected when modeling problems in Nature. The impossibility until relatively recently, *i.e.*, the last third of the 20th century, of computing such representations efficiently, or at all "within anyone's lifetime" for that matter, and their inherently inelegant appearance as represented by the system of notation available, doubtless drove many away from even considering these phenomena as worthwhile subjects of investigation. Many researchers applying mathematics to modelling real-world phenomena would likely reject, as an extremist position, the militant insistence of the British mathematician G.H. Hardy (1940), briefly the mentor of the Indian mathematical genius Ramanujan, that one should approach and present mathematics as some kind of pure thought-experiment continuous in time, untainted by (and having nothing to do with) any possible application. Nevertheless, Eurocentric conceptions, stemming from ancient Greek philosophy, of beauty as a function of two-dimensional symmetry, "balance", etc. remain very much part of the expectation of most mathematicians alive and working today on current problems of both pure and applied mathematics. This has served to reinforce a tendency to discard or dismiss as unlikely an "inelegant-looking" result.

Chapter 3

FROM ILLUSIONS OF PRECISION AND REPRODUCIBILITY IN NATURAL SCIENCE TO DELUSIONS OF NORMALCY IN SOCIAL SCIENCE

Newtonian calculus had become cluttered with refinements and special recipes of all kinds by the 19th century. The precision and especially the reproducibility of results achieved using it were nevertheless remarkable. The physical sciences were written about and spoken of as "exact sciences". There were not a few who understood very well the price of such progress. Appropriate initial and-or boundary conditions had to be established in which a given differential equation could be applied. An inappropriate selection would render any results from using the equation meaningless. There were probably rather fewer who also understood that preparatory research would be required: *before* selecting and applying any existing linearising model equation to the task of extracting possible solutions, exactly how invariant any actual initial and-or boundary conditions might be, with time, for a process taking place in nature has to be established. Against this overwhelming current, however, who was going to look back and question the applicability to the reality of nature of methods and models emerging from the linearising assumptions of real analysis? Unfortunately, this set the context in which researchers in social science also became concerned with rigourising their methods in the middle of the nineteenth century. The rigourisers were feeling increasingly pressured from two directions.

Dynamism is inherent in all social or individual development. The idea that equilibrium is normal and anything other than the steady state is a disturbance and disruption is a notion that has served every Establishment in all times and places.

However, it is not a true description of social, economic or political reality. Such striving for the steady state emerged clearly at the time of Sir Isaac Newton. It invaded and permeated his scientific work. From his time to date it seems hardly accidental that "success" in a scientific career is correlated strongly with supporting the *status quo* of the day. Is this an accident, or a position based on prejudice rather than science? Certainly the historical argument can be made that, in Europe during the Inquisition and earlier, scientific integrity pitted many researchers openly against the authority asserted over scientific matters by the Church, with Galileo representing only the most dramatic and highest profile in a long line of similar cases that preceded him.

The argument can be advanced even more compellingly on scientific grounds. If such a thing as steady-state equilibrium is possible, and actual, anywhere in Nature, how is it *also* possible that matter and energy can be neither be created nor destroyed, but only change form, sometimes even changing one into the other? One or the other: *either* steady state, in which case neither matter nor energy can be changing form, *or* motion is the mode of existence of matter. This is fudged in various ways. For example, repetitive forms like reciprocal or cyclical motion are often represented as a kind of pseudo-steady state within a delimited range. However, the maintenance of real-life reciprocal motion, like that of pistons in an internal combustion engine, requires a directed expenditure of energy in a bounded chamber that ceases once the supply of combustible fuel is cut (by running out of fuel or turning the engine off). This is a human-engineered phenomenon, normally not found anywhere in Nature. Cyclical repetition in Nature does not repeat the exact same path in each circuit, any more than the Earth repeats the identical path in its orbit around the sun. Even the repetitive cycles of "chaotic attractors" (like Julia or Mandelbrot sets) generate an infinite number of "self-similar" but unique, non-identical cycles.

The reality – that, regardless of what can be engineered to happen for some finite period, there exists no such thing anywhere in Nature as a steady state – has been masked. Instead Newton's First Law of Motion is widely accepted as the first and last word on the inertial properties of *matter*. This provides that "an *object* at rest tends to stay at rest and an *object* in motion tends to stay in motion with the same speed and in the same direction unless acted upon by an unbalanced force." On its face, this law does indeed appear to provide definitive criteria for the analysis of inertia in all possible cases – at rest, or already in motion. However, in fact, it is at the very least a potential source of disinformation because 1. *resistance to* motion is identified, at the empirical level of "objects", with *absence of* motion, and 2. even at apparent equilibrium, *i.e.*, at a point between a previous completed state of motion and a pending resumption of motion at some subsequent stage, something at the microscopic level of matter remains in motion, *e.g.*, at the molecular level. If Newton's First Law is loosely applied to all forms of matter in general, however, motion ceases to be an inherent property of matter. Once this separation is effected, all kinds of mischief comes in its wake.

A great paradox and howling contradiction is widely accepted without further thought. is that, on the one hand, we cannot have motion without equilibrium moments, whereas the disinformation is that all motion tends towards equilibrium in the presence of an appropriate balance of forces. In the unceasingly dynamic environment that continues to exist outside and around a stationary observer, how matters may appear to such an observer can hardly be accepted as the final word or definitive description of what is actually taking place. In other academic realms apparently far beyond or unrelated to natural science, such as politics and social science, the same lure of the steady state is undeniable, but once again its aphenomenality inevitably leaps out. Consider Georg Wilhelm Friedrich Hegel (1833), the philosopher of the modern era credited with doing more that anyone since the ancient Greeks to restore respectability to the notion popularised by Heraclitus that "all is in flux". Hegel wrote that the autocratic, militarily powerful, small German-speaking state of Prussia in the 1830s represented the apex of human achievement in statecraft, and a balanced polity which could not and would not be further improved upon, *i.e.*, nothing in Prussia was any longer in flux![1]

The first challenge posed in the modern era to the staeady-state paradigm came in the works of Karl Marx (1867). It was widely followed up in many others' work by the start of the twentieth century. According to this new paradigm, the entire social order currently and historically never represented any such thing as true equilibrium or steady state. The bourgeois order of 19th century Europe was dominated by the societal model of France and the industrial model of Great Britain. Many viewed this order as the epitome of "Progress"-in-general

---

[1] Following the implosion of the Soviet Union, the former U.S. State Department official Francis Fukuyama (1992) famously declared "the end of history", meaning the U.S. system had now demonstrated its superiority forever to the end of what remained of recorded time. In 2003 Fukuyama declared it was the destiny of the United States as the premier upholder of the banner of freedom and democracy on earth to invade Iraq and topple Saddam Hussein, but in early 2006 recanted by declaring the entire affair a great "mistake". These prognostications reflected something other than a scientific investigation of historical developments and their direction or meaning: the Prussian government paid Hegel's salary as a university lecturer, while the network of opinion-makers and government officials for whom Fukuyama was an occasional mouthpiece went from riding high in the opinion polls and counsels of the executive branch to losing the public's trust. If dynamism is Nature's reality, denial of it in any form or to any degree is always part of the agenda of some force or other from the Establishment be it in government or academia.

with a capital "P". For them, this message of incessant societal flux was anathema. Influenced by the positivist trend pioneered in French philosophy and social science by Auguste Comte (1848), many researchers in social science saw development in their own time as a struggle between "forces of progress" and forces opposing progress. The present moment in western European social development was identified as "progress." All opposition was portrayed as potentially or actually opposed to progress. All tendencies reinforcing the current line of development within the status quo, especially everything tending towards equilibrium, were presented as a support for progress (Butterfield 1968). In this way, time based on the steady state, *viz.*, an inherently linearised conception of time, was confounded with the notion of time as a measure of "progress." Subsumed by the achievement of equilibrium, the irreversibility of actual time was made to disappear.

INTANGIBLE	TANGIBLE
Value *(Value-in-Use)*	**Price** *(Value-in-Exchange)*
Surplus	**Profit**
TIME (Historical)	**TIME (Linear, i.e., t2 – t1)**

Figure 2-5. Tabular comparison of some tangible and intangible categories of economics.

The Marxian categories of "value" and "surplus" included the more familiar notions of "price" and "profit, respectively, as tangible subcategories. The concepts of time associated with the intangible and the tangible sets of categories were $t_{HISTORICAL}$ and t_{LINEAR} respectively. See further discussion at *2.3.2* below.

Thus was the notion of $t_{NATURAL}$ extended now to include social phenomena – but t_{LINEAR} and $t_{NATURAL}$ still remained one and the same.

Furthermore, according to this logic, any disturbance of this equilibrium was illegitimate. Such disturbances were to be accounted as the work of deranged, deviant, alien sources and forces. The French sociologist Emil Durkheim acknowledged at the end of the 19th century that the society itself could be the seeds of many of these disturbing phenomena, but the individual was ultimately responsible in any particular case since deviance itself manifested itself individually. (Durkheim 1897)

Long-term thinking empowers people and enables people to empower themselves. (The moment one becomes hooked on long-term thinking as a habit, it becomes blindingly clear that all true empowerment derivesa from oneself and one's relations with other people, not from any externally-imposed or externally-induced condition.) Long-term thinking means thinking about the consequences of one's own actions. It also means re-examining everything reported from the standpoint of extracting its longer-term significance, beyond what is being immediately reported. Within the very process of daily living, it ought to be well within everyone's interest to apply long-term thinking in all times and places. Yet, this has not happened. There are what are called "pressures of daily life" – which actually means pressure to produce some outcome in the short term – which are usually blamed for this shortfall. But the fact remains, and it cannot be made to disappear by being glossed over, that it is actually in the *interest* of the vast majority of individuals, viewed either in the short term or for the long term, to apply long-term thinking as a habit in all times and places. Human conscience exists in all times and places and it will always assert its claims in the field of human action. Just because individuals can and do frequently suspend listening to their conscience does not make it go away or disappear. The key to maintaining long-term thinking is to suspend listening to, or being pressured by, anyone and anything that places the interests and needs of the short-term ahead of the long term. If the long term is not continually attended to, there will be not only no long term but the short term will become far shorter. There is nothing at all mysterious about long-term thinking. Start with clarifying *where* whatever it is you are thinking about fits or exists with respect to the Past, the Present and-or the Future, and with *why* it is of any significance or importance to you. That already takes care of two profoundly significant intangibles: time and intention. Just as there cannot be such thing as matter without motion (*i.e.*, energy), there is no such thing as understanding, *i.e.*, meaningful knowledge, without the individual taking action to "find out". A positive, *i.e.*, pro-social and long-term, intention ensures that the seeker will find something useful. Finding out something for oneself is the most empowering thing there is. "Learning" something on the say-so of some "authority" is the most disempowering and enslaving thing there is. What is so especially empowering about long-term thinking is not whether this-or-that piece of knowledge was already known or even previously thought about, but rather the journey on which it takes the seeker.

An associated concern expressed by many of the "rigourisers" about Marxist approaches had to do with the Marxians' foundation in historical scholarship. The data of history disclose patterns that could be explained in terms of the actual social and economic structures in place at the time. However, Alfred Marshall, the founder of neoclassical economics as an academic discipline, argued that such historical consideration and analysis were useless and irrelevant anywhere in the social sciences. In his view, Darwin's theory of evolution demonstrated that the only factors decisive in any process of change would be found among those most recently-generated, not among those historically handed down (Marshall 1890). Here lies the source of two unwarranted assumptions: 1. the closer time t is to "right now", the less-qualified and more precise will be the mathematical rendering of whatever the social condition being studied, and 2. the justification for resorting to steady-state, equilibrium-based models. Although Marshall may have believed his theories were objective to the extent that they hewed to this notion of the relative indifference of temporal factors, his prejudice itself had little to do with the correctness or incorrectness of his own theory in particular. The late John Maynard Lord Keynes was probably the greatest refuter among 20th century economists of much of the *corpus* of Marshall's neo-classical assumptions and analysis. He also believed that historical time had nothing to do with establishing the truth or falsehood of economic doctrine. "In the long run, we are all dead," he wrote. He tied this to a stance that attacked all easy acceptance without question of any of the underlying assumptions propping up all forms of orthodoxy. Accordingly, this retort was taken as the sign of a fresh and rebellious spirit. However, in his own theoretical work he was frequently at pains to differentiate what happens to individuals who are driven by short-term considerations from what happens at the societal level at which he was theorising about broad historically sweeping movements of economic cause and effect (Keynes 1936).

Although not alone in this interpretation – many of his peers in both the social and natural sciences shared it – Marshall for his own purposes seriously misconstrued the thrust of Darwin's argument. Darwin said only that the emergence of a species distinct in definite ways from its immediate predecessor and new to the surrounding natural environment generally marked the final change in the sequence of steps in an evolutionary process. The essence of his argument concerned the non-linearity of the final step, the leap from what was formerly one species to distinctly another species. What Marshall had in mind *viz.*, the proximity of that last step to the vantage-point of the observer – which might be centuries or millennia or longer – was *not* the relevant temporal factor. Rather the length of time that may have passed between the last observed change in a species-line and the point in time at which its immediate predecessor emerged – the characteristic time of the predecessor species – was the time period in which all the changes so significant for later on were prepared. This latter could be eons, spanning perhaps several geological eras. This idea of $t_{NATURAL}$ as characteristic time was the one which Marshall's misappropriation of Darwin's idea obscured. Even though Keynes was prepared to suspend obsessions with the short-term for the purposes of establishing his analysis of the broad historical sweep, he stood side by side with Marshall in marginalising any role for or consideration of $t_{NATURAL}$ in the setting of actual policy, especially where economic gain could be enhanced in the short term by disregarding it.

A second, but equally telling source of pressure on social scientists to mathematise their research methodology was a sense that their work would not be taken seriously as scientific without some such mathematical rigour. As the models and mathematics from the so-called "exact" sciences would hardly be appropriate or seem credible in any field of study focusing on human beings and their incredible variety of needs, wants and impulses, another kind of mathematics would have to do. Questions of history and historical phenomena were also a convenient target because of the lack of any means to describe them with any meaningful, non-trivial mathematical model.

The solution to both the Marxian-influenced challenge concerning incorporation of historical time-scales and the "rigourising" challenge was to come from ... statistical science, based on theories of mathematical probability and its probability measures of "uncertainty". The disinformational role of statistical and probabilistic models in the social sciences was discussed earlier in the section on "laws of motion", "natural law" and questions of mutability. A particular area of interest in connection with the concerns about "rigour", however, was that part of mathematical statistics based on the theories of probability in which the discrete is analysed on a scale in which it could be said to approximate the continuous. This maneuvering around the discreteness of social events finessed the entire question of discontinuities in general, including the discontinuity that marks the onset of some new turbulence. (There will be frequent occasions throughout the present volume to comment on and mark some of the most egregious cases of this maneuver). The underlying probability distributions for these "continuous approximations of the discrete" were exponential mathematical functions, thereby furthering the tendency to linearise.

The current state of affairs has made it very difficult to remember, and appreciate the consequences of, a quite fundamental fact: the mathematical models used and applied to "get a handle" on engineered and-or natural phenomena are, first and foremost, *images* of an ideal form. Today, when it comes to appreciating scientifically and correctly the intangible aspects of temporal factors of any phenomenon, scientific workers – having been put through the ringer of quantum mechanics, etc. – know that t_{LINEAR} and $t_{NATURAL}$ cannot possibly remain one and the same. Although a linear model applied to linearly-engineered phenomena is known to work well under circumstances where some operating limits have been experimentally verified, a linear model applied to phenomena that are themselves not linear is another matter.

Nothing in nature is linear. Linear independence cannot be a feature of any model purporting to reflect the reality of a situation where everything affects everything else and there are literally dependencies upon dependencies; the notion of any system operating in isolation, or of modeling the solution of any problem presented in Nature by assuming the condition of some isolated system or sequence of such systems, is aphenomenal. Changes of state occur, appear or disappear both continuously as well as discontinuously. There is no such thing as "steady state". Problems as found in their natural setting always appear "ill-posed" but up to now there seems to have been a concerted effort not to attempt solutions to problems in this state. Instead a problem that looks like the actual problem but which can be posed in more or less linear form is solved instead and this result is declared to be something approximating the solution of the actual problem... given the addition of certain conditions and boundaries to the original problem's definition. This distorting technique starts very early, with high-school/first-year university instruction in how to solve problems associated with simple harmonic motion of a pendulum using Newton's Second Law of Motion in a linearised approximation as the governing equation. As mentioned above, one convention widely adopted up to now involves artificially and arbitrarily hedging the reality to be observed in nature with various time-constraints so that some relatively tractable mathematical model may be applied. This is not unrelated to the fact that the solution schema developed for such models have become ever more elaborate. These linearised images have served to sustain an illusion that nature's secrets are being discovered at a rate that is in lock-step with the advances taking place in the technology of electronic computation. Somewhere, somehow, no matter how general and even non-linear the governing partial differential equation to be modeled and solved, linearisation continues – a truth reflected either in the ongoing production of unique solutions, or small sets of multiple solutions, from these models (Islam 2005).

This is a final and total abandonment of any last shred of scientific integrity. To seriously propose such solutions and models is to proceed utterly deaf to the ancient injunction that "Nature cannot be fooled" (Feynman 1988). Over the past more than 300 years going down Newton's road, these are the typical consequences that flowed from the more-or-less uncritical assumption of the difference-quotient formula for the derivative, premised on evaluating "in the limit", *i.e.*, as $\Delta t \rightarrow 0$.

Time rendered tangible by Newtonian linearisation proceeds at a pace whose uniformity or otherwise is a matter of indifference. For many real-life situations, such a misrepresentation of an intangible of such importance has frequently produced meaningless results, sometimes leading to results so ludicrous that "interpolation" was required. Is the Newtonian linearisation all there is? One meaningful and non-trivial mathematisable conception of "historical time" might be to consider the passages of time in terms of cycles whose periodic features change at specific branch-points. This would entail a complete and final break with any notion time linearised by Newtonian methods. "Well-behaved" functions whose initial and-or boundary conditions permit extraction of a solution or solutions would be subordinated to the reality of mathematical chaos, where extreme sensitivity to initial conditions carries "stability" implications about the meaning or existence of a function over one or more sub-intervals nested within the function's overall domain of interest. In addition to restoring a modicum of humility as well as integrity to the scientific enterprise by solidly repositioning the non-linear as the general case and the linear as the exception, such an approach should make it possible to disclose and elaborate the "characteristic time" of many real-life processes in nature (Islam: *ibid*). In this connection, Chapter Six addresses the possible rationales for encouraging investment for periods exceeding the expected human lifespan, as well as for recalculating "return on investment" in terms of what would be a meaningful characteristic time of project's actual service to Humanity. An important consideration is that the timespan adopted for purposes of analysis exceed whatever the characteristic lifespan.

Chapter 4

MUTABILITY

4.1. "LAWS OF MOTION", "NATURAL LAW" AND QUESTIONS OF MUTABILITY

In science, not all laws are equal – some may describe an empirical relationship, others define fundamental features common to an entire category of processes. Newton's calculus held out a seductive promise of all relationships becoming in principle quantifiable, even computable. However like all "law", this promise was actually a double-edged sword. Would such laws as those that Newton's calculus might describe be relationships that captured the reality of change in the natural world, or would they be mere snapshots freezing some relationship in an artificial bubble of permanence?

It is entirely possible for quite fundamental laws to operate even as their very existence remains vehemently denied. Take, for example, the basic law of operation of capitalist political economy, which is that the rich get richer and the poor get poorer. On the other hand the basic law of operation for the development of theoretical models of this political economy is to stop at nothing to deny or obfuscate this basic law of operation of that actual economy.

A recent interview by the London *Independent* (Vallely 2006) cast 2001 Nobel Economics Laureate Joseph E. Stiglitz as an Old Testament prophet. The tradition of religious prophets was that they pinpointed defects in the current course of their societies on the basis of disclosing the truth, measured against an unvarying moral standard, about the society's past, present and future as they had come to understand from observing contemporary events unfold around them. Stiglitz's own words as relayed by this interviewer seem, however, to disqualify him from membership in such an exalted category. For example, he is quoted in the following context saying global warming shifted within a decade from being a theoretical possibility to a serious threat today (all emphases added):

> A decade ago Mr Stiglitz was a member of the Intergovernmental Panel on Climate Change. Today his concern about global warming has been turned into alarm. *"Ten years back the theory was clear, as was the evidence of the increasing concentrations. But no one thought it would manifest itself as quickly, and in such a dramatic way."* (Vallely: *ibid.*)

What has accumulated in the last decade is a massive amount of data that could be evidence of... just about anything, from climate change on a cycle as long as several human lifetimes to planetary warming whose reversibility is unknown (because the only projections that have been investigated are cascades of consequences, rather than any set of indicators of a temporary equilibrium in the future that will differ from whatever temporary equilibrium has been put behind us by developments of the last two generations). While the increased presence of CO_2, an essential component of the atmosphere that makes human and other life possible on this planet, has been identified, no differentiation has been made between fresh and extremely old CO_2, and little has been elaborated concerning the pathways of the vast number of toxic byproducts of petroleum refining and other chemical processing borne aloft with the CO_2. To suggest that his "concern... has been turned into alarm" by what has happened over the last decade is to engage in disinformation. If someone tells me: "I am more scared today by what I don't know than by what I dismissed 10 years as relatively unimportant in the near term", my instinct is to reply: "Then stop playing 'Chicken Little' waiting for the sky to fall and go find out!" But without addressing the business of actually finding out the science of climate change, Humanity is indeed paralysed. According to the interview, however, without doing anything to establish the science of anything, the problems that now loom so large in the short-term that we can no longer even discuss the possibility of a future can be solved by a couple of quick fixes:

> Mr Stiglitz offers two solutions *[to managing global warming - Ed.]*. The first is to increase incentives for developing countries to get involved in global warming reductions. Carbon-trading initiatives offer a market-based solution. But there is need for more. "Kyoto offered financial rewards to Third World countries for planting new forests, but not for maintaining existing ones. So Papua New Guinea can get money if it chops down its forest and replants it but not if it just keeps its forest. *That's silly.*" (Vallely: *ibid.*)

What the interview report leaves out is that developed countries with large forest cover, including the U.S., Canada and post-Soviet Russia, are indeed permitted not only to count such forests but to trade them with developing countries for emission credits – that's not silly, that's dead serious – but neither Brazil nor countries to its north and west that share any portion of the Amazon basin are permitted to count much less trade for emission credits any part of the Amazon rain forest under the pretext that the Amazon rain forest forms part of the common heritage of Mankind and is not for sale! The disinforming, paralysing part is that, by insinuating that independent development in developing countries outside the Kyoto framework would at least compromise if not threaten outright the ability of Humanity to breathe clean air and drink fresh water, this position also simultaneously pre-empts and forecloses such an alternative. This report creates that disinformation by failing to point out that emission-trading schemes are premised on developed countries being able to buy or to sell these credits, while developing countries are permitted only to buy credits, and even then only from countries in the developed-country group. What Stiglitz accurately describes as a "market-based solution" is taking place, but for the purpose of preserving an economic environment in which developed countries continue to call the tune in developing countries. However, when it comes to establishing the real test of prophecy these days, one's positions on global warming and Kyoto have been completely displaced by the matter of where people stood at the time the U.S. government began publicly declaring that Saddam Hussein was so dangerous that only a full-scale invasion could save Mankind from the fate his continued presidency in Iraq held in store. Tens of millions of people participated in protest marches in almost 3,000 cities and localities around the globe in the four months preceding the invasion: clear, one would think. But the *Independent* interview quotes Stiglitz on this question as follows:

> "..It's hard to rebuild infrastructure *[in Iraq - Ed.]* because of the insurgency. It's hard to deal with the insurgency because of lack of jobs. And it's hard to provide jobs because of lack of infrastructure.
>
> "Those, like me, who warned they were walking into a quagmire were more right than even we supposed. The lack of analysis and preparedness of the Bush administration - and of Tony Blair by association - was astounding, particularly since this was a war of choice .." (Vallely: *ibid.*)

If this indeed be the test or proof of prophecy, or of qualifications to be considered an "Old Testament prophet," perhaps it is time to re-examine that very notion and the ancient traditions associated therewith. Parodying Groucho Marx:

the others might no longer wish to be associated with a club that could accept Joseph Stiglitz as a member.

This is an all-sided continuous campaign. It began long before Marx and the publication of *Capital*. Since then it has become far broader and more desperate. Its aim, however, remains the same: to ensure survival of the status quo by stifling any consciousness or source of consciousness about any alternatives. As one of the recent leaders of this campaign, former British prime minister (now Baroness) Thatcher, used to intone: "*There Is No Alternative.*"

Using the initial letter of each word in the phrase to form an acronym, critics have labelled this ongoing campaign the "TINA syndrome". Over the 25 years or so, this campaign has emerged in a wide range of manifestations, throughout all fields of study in politics, economics and policy. Throughout the social sciences and even in the natural sciences, assertion of the TINA syndrome and the struggle against its assertion have together spurred an intense and renewed interest in the meaning of "law" in general, and of how particular processes may be considered to be governed by some sort of law, or pattern, or set of relationships. It is difficult enough to conceive anything more intangible than a "relate" or relationship, let alone one such as the TINA syndrome that has produced such wide and highly tangible impacts. There is indeed no alternative at this point but to take the plunge and examine what the brouhaha is all about.

The industrial revolution was already under way for a generation in Britain when Adam Smith famously put forward his theory of the so-called "invisible hand":

> ..every individual necessarily labours to render the annual revenue of the society as great as he can. He generally, indeed, neither intends to promote the public interest, nor knows how much he is promoting it. By preferring the support of domestic to that of foreign industry, he intends only his own security; and by directing that industry in such a manner as its produce may be of the greatest value, he intends only his own gain, *and he is in this, as in many other cases, led by an invisible hand to promote an end which was no part of his intention.* Nor is it always the worse for the society that it was no part of it. By pursuing his own interest he frequently promotes that of the society more effectually than when he really intends to promote it. I have never known much good done by those who affected to trade for the public good.
>
> Adam Smith, *An Inquiry into the Nature and Causes of the Wealth of Nations* (1776) [Emphasis added – Ed.]

Implicit in Smith's invocation of the superiority of the individual pursuing his self-interest over the interests of society or the public lies a notion of the shortest conceivable time-span, one in which $\Delta t \to 0$: "he intends only his own gain". Chapter Five discusses this aspect of the aphenomenal model – it is actually a defining feature – by considering what happens to "self-interest" transplanted to a context in which $\Delta t \to \infty$: clearly, self-interest in the long-term becomes the pursuit of gain or benefit for society as a whole. Otherwise, it would be akin to dividing by zero, something that would cause the model to "blow up".

All the defenders of, and apologists for, the status quo have pointed to Adam Smith's argument as their theoretical justification for opposing in principle any state intervention in the economy. Chanting their mantra of the invisible hand, policy-makers on the same wavelength have been confining and restricting such intervention to those parts of economic space in which there operates no profitable production of goods or services with which such intervention would be competing. At the practical level, the only question remaining about this "invisible hand" is whether its invisibility arises from absence, *i.e.*, non-existence, or from darkness, *i.e.*, a sinister existence like the Black Hand.[2]

One rather profound and deeply disturbing truth about the TINA syndrome that has begun to dawn far more widely than ever before since the disappearance of the old Soviet bloc is that neither the State as bogeyman nor the State as employer-substitute is a viable option for Humanity. There is in fact no alternative but something other than either of these options. This is a subversive consciousness very much resisted by both proponents of TINA and their detractors. Once $\Delta t >$ some characteristic time, and a measure of the change in economic space, s, exceeds the lone individual, *i.e.*, as $\Delta s >> 1$, then the role of the human factor – social consciousness can indeed become decisive. It increasingly must displace the need for a State to play all or any of its previously accustomed roles in the economic life of society. At that point, intentions that serve the society as a whole no longer require the application of some power previously delegated to an external force (the State) in order to prevail as a norm. That is the stage in which the intangible – good intentions – can finally command the tangible.

At the theoretical level, the significance of Smith's observation of the so-called "invisible hand" is that the outcome of normal operations of commodity production are achieved independently of the will of any individual participant or group, *viz.*, "an end which was no part of his intention".

---

[2] The Black Hand was the Serbian secret society, likely modelled on a much older secret society of the same name founded in Sicily in the 1400s and said to be the origin of the Mafia, which carried out the assassination of the Austrian Archduke Ferdinand in June 1914, precipitating the outbreak of WW1. The Serbian Black Hand's history remains so tangled that to this day no authority is absolutely certain what combination of European intelligence services actually financed the crime. Ever since, the Black Hand has become a metaphor for the murkiest of murderous mayhem (Tuchman 1962).

Significantly, Smith does not say that money-capital wedded to the short-term immediate intentions of some individual (or grouping of common interests) will not achieve its aims. Rather he confines himself to observing that objectives which formed no part of the originating set of immediate short-term intentions, *viz.*, "an end which was no part of his intention", may also come to be realised thanks to the intervention of the "invisible hand". Chapter 6 will discuss what can happen with intentions in our own day by applying the economic theory of intangibles advanced in this book, so that intentions are translated and expressed in a socially positive manner, for long-term aims.

Smith believed that "an end which was no part of his intention" came about as a byproduct of how competition operates to "regulate", in a rough and overall manner, both the supply of and demand for socially necessary goods and services. The will of any consumer(s) or producer(s) by itself would never suffice. The secret to the "law of motion" of an industrial commodity economy lay in how the marketplace under conditions of free competition allocated economic resources.

Underlying Smith's view was a broader 18th-century Deist philosophical outlook already prevalent among a broad section of the European intelligentsia of his day. Anything could be examined as the outcome of a process comprising observable, definable stages and steps, and linked ultimately to some Prime Mover (or initiating force). The scientific model for such narratives was provided by Sir Isaac Newton's *Principia Mathematica*, crowned by his discovery and elaboration of the laws of motion and the principle of universal gravitation. (Newton 1687)

For most scientists of the 17th and 18th centuries, an analysis ascribing a process to some Prime Mover manifesting itself as Newtonian "mechanism" was the best of all possible worlds. On the one hand, a natural occurrence could be accounted for on its own terms, without having to invoke any mystical forces, divine interventions or anything else not actually observed or observable. On the other hand, the divinity of Creation need not be dispensed with or challenged. On the contrary: this divinity was being reaffirmed, albeit indirectly "at a certain remove" insofar as whatever was required to sustain or reproduce the process in question could now be attributed to some even more fundamental "law of motion".

The revolution occasioned in scientific outlook since the publication of Charles Darwin's *Origin of Species* (1859) has become so complete and all-encompassing that, from the vantage point of the start of the 21st century, it is hard to remember that much of the support for, and embrace of, Newtonian mechanism (and the attendant penchant in many fields for "laws of motion") derived from the belief that it could be reconciled with a Creationist assumption, not just about Man *within* Nature, but about the very existence of Nature itself.

Re-examined in this light, the impact of Smith's assertions about the "invisible hand" among his contemporaries can be better understood. In essence, he was declaring that economic life comprised phenomena that could be analysed and comprehended as scientifically and as objectively as Newton had analysed and disclosed the laws of physical motion of all forms of matter in Nature and even the universe. Furthermore, such investigations would provide yet another proof of the divinity of Man's existence within that natural universe.

Between the time of Sir Isaac Newton in the early 1700s and that of Charles Darwin in the middle third of the 1800s, these considerations were framed and understood by scientific investigators within a larger context, *viz.*, the conception of "natural law". Using scientific method, Man could come to know, understand and make use of natural laws – laws operating within observable processes in Nature itself, and discoverable from systematic observation of these processes. However: *these natural laws in themselves were immutable.* This was the same as with any mathematical function whose "Newtonian" derivative yielded an instantaneous rate of change between points on its graph but which itself did not change. In fact, it was precisely this notion of the immutability of natural law that was assumed and implicit within the general and more widely-accepted view that some law of motion, eventually connectible back to a Prime Mover, must account for any and every process observed in Nature.

The conundrum was simply this: if natural law were not immutable, science would be compelled to account for innumerable random divine interventions in any natural process, at any time. Such a course could drag science back into the swamp of the metaphysical idealism of Bishop Berkeley – Newton's great antagonist – who famously explained that objects in physical nature continued to exist beyond out perception because God exists to cognise them while and whenever human beings are not available to cognise them (Berkeley 1734). In the words of a limerick popularised widely in the 19th century specifically satirising Berkeley:

> *There was a young man who said "God*
> *Must think it exceedingly odd*
> *If he finds that this tree*
> *Continues to be*
> *When there's no one about in the Quad."*
>
> *"Dear Sir, your astonishment's odd;*

> *I am always about in the Quad*
> *And that's why this tree*
> *Will continue to be*
> *Since observed by Yours faithfully, God."*

No one would accept something so contrary to common sense; science and scientists would be come laughing-stocks. If natural laws were not held to be immutable, how could logical reasoning guarantee that error could be detected and rejected?

The actual solution of this conundrum in practice came in the course of further, deeper-going research into actual phenomena. Since the middle of the 19th century, starting with Marx in social science and Darwin in natural science, and extending early in the 20th century to physics and chemistry with the elaboration of theories of quantum mechanics, it has become increasingly clear that the mutability or immutability of any natural law is actually a function of, and dependent on, the time-scale selected for observation and study. The problem here in general is one of method. The particular source of the problem lies how the methods of scientific investigation that are applied to comprehend the material deal with temporal factors, the passage of time, the role of time.

In social science, the appropriate time-scale is the historical period of a given social mode of production. Before the epoch of a given mode of production, for example, there might be a certain law of population growth/decline, but with the emergence of the new epoch this law would change its form and-or manifestation. Thus Rev. Thomas Malthus' notorious extrapolation of population growth overwhelming increases in food production depended on a failure to distinguish, on the one hand, the disappearance – in less than two generations, as people left to find work in the new industrial centres – of a rural population that had been stationary for the preceding five and one-half centuries from the rapid increase, on the other hand, of population in the industrial centres during the same period. The law of population growth/decline governing the latter was bound to be entirely different from the law of population growth/decline governing the former because the manner in which the society procured its food supply and other needs had been completely transformed. By ignoring this distinction, however, Malthus' *Essay on Population* (1798) perpetrated a misunderstanding that persists to this day with periodic predictions reappearing from time to time of global collapse due to overpopulation, regardless of the fact – thoroughly and repeatedly exposed by Boyd-Orr *et al.* (1937; 1940; 1943) – that not a single one of any of the hundreds of similar such predictions in earlier periods has ever been validated by events.

In the geological record, entire species appear in one epoch only to disappear in a later one; ludicrously, this has been adduced by so-called "Creationists" as evidence that Darwin's theory of evolution – which used such leaps and gaps precisely to explain speciation – must be untrue! Of course, evidence of this kind proved only that the notion that evolution should take place as a smooth process uninterrupted by quantum leaps – the very view that Darwin's analysis and evidence definitively refuted – was devoid of reality. The same issue of time-scale is now just beginning to be understood regarding some of the earliest states of matter in the first few picoseconds of the Big Bang.

With the exposure of these absurdities, it has become possible to start hammering the final nails into the coffin of the "TINA syndrome". All phenomena or effects duly observed in any natural or social process arise from some verifiable cause, but in accordance with the operation of some body of law that remains constant and consistent, and always within some definite spatio-temporal boundaries. To argue immutability outside such boundaries is open to serious question, while to deny or ignore the existence, and the role or consequences, of such boundaries is a source of scientific disinformation.

"Scientific disinformation" is a most apt description of the condition in which provision of scientific theory and researched data nevertheless leave the social order incapacitated when it comes to framing and-or selecting a course of action to carry out consciously programmed changes in the status quo. It explains very well why, for example, literally millions of people in our own time have become perfectly well aware that the existing social and economic system itself reproduces a condition alluded to at the start of this section – in which the rich get richer and the poor get poorer – but nevertheless no agency of the social order is capable of intervening to turn the situation around.

Assume for the moment that this societal condition is recognised as a scientific and verifiable fact. For the moral philosopher, the matter of responsibility is immediately posed. For the more dispassionate scientific observer, it would be important to establish causes and effects in order to sort out the dynamics of this condition. How to alleviate the negative consequences of such a condition in various areas – the health of the population, the education of the upcoming generation, etc. – would accordingly preoccupy specialists in the relevant respective fields of social science and policy. However, there is indeed a way to present the evidence of this condition in its various aspects, and of the extremely negative consequences flowing from this condition, so that everything is to blame for the condition, and hence no one thing is to blame for any part within the overall. One approach that fills this bill very nicely is the resort to statistical methods in social science – especially those involving correlation.

One of the most important consequences of resorting to statistical methods was the finessing of the need to establish and distinguish cause from effect. To be able to assert that A and B are related by some correlation coefficient χ appears highly suggestive of underlying reality even as it skirts at the same time the entire issue of whether $A \rightarrow B$, $B \rightarrow A$, or actually $Q \rightarrow A$ and $R \rightarrow B$ while in fact no causal relationship whatever exists between A and B. Correlation is very useful where causal relations are already known and established. In the social sciences, however, in the absence of – or inability to gather – any other evidence from more direct or more thorough experimental observation, it has become *de rigeur* to employ correlation to imply or suggest a causal relationship. Is the publication of caveats about the distinction between demonstrating a correlation and suggesting some relationship of cause-and-effect sufficient to shield such activity from merited condemnation as a serious abuse of the requirements of scientific integrity?

One of the most fundamental requirements of science properly conducted is that one's work at the end of the day draws some line of demarcation between what is known to be false and what may not yet be fully understood to be the truth. Detection of error and elimination of falsehood are absolutely fundamental to scientific enterprise at any level. In this respect, the "Correlation" bucket has holes in it big enough for a veritable spotlight to coruscate. Consider the following example. If one were to correlate "intensity of religious faith", "presence of exact bus fare" and "frequency of arrival at a preset destination on public transit", any number of clearly nonsensical, as well as a number of apparently reasonable, correlations might be elaborated, *e.g.*, "faith and a two-dollar coin gets you downtown on the bus." However, regardless of how anyone might go about weighing the various possible renderings of the available evidence, the results would always be insufficient to rule out possibilities lying on the farthest margins and perhaps bordering on nonsense, *e.g.*, what happens if you have the two-dollar coin but lack faith? This converts the likely acceptance of the apparently more reasonable-seeming possibility (or possibilities) into a matter of purely personal prejudice. It is no longer guided by a procedure that meets the fundamental requirement of any scientific method, *viz.*, that a clearly erroneous result will be excluded by the weight of the evidence and not by the prejudice of the investigator.

Statistical modes of reasoning carefully employed, in a context where there exists some actual knowledge of definite causes and definite effects, can be subtly powerful. But it is an entirely different story when reasoning proceeds from the grouping of data according to statistical procedures derived from the norms of some abstract probability distribution. No groupings of data, however well-fitted to some known probability distribution, can ever be the substitute for establishing actual causes and actual effects. Substitution of the "statistically likely" or "probable" in the absence of knowledge of what is actually the case is a truly inexcusable breach of scientific integrity. For one thing, speaking purely in terms of how the logic of an explanation for a phenomenon comes to be constructed when inputs are "probable" or "likely" but not actually known, if any of the steps on the path of reasoning toward an actually correct conclusion are themselves false, neither Bayesian methods of inferring conditional probabilities (Jevons 1870) nor Pearsonian methods of statistical correlation (Pearson 1892) will assist the investigator to reason to the particular conclusion that will be demonstrably most consistent with known fact. Consider this syllogism:

– All Americans speak French [major premise]
– Jacques Chirac is an American [minor premise]
– Therefore Jacques Chirac speaks French [conclusion-deduction]

If the information relayed above in either the major or minor premise is derived from a scenario of what is merely probable (as distinct from what is actually known), the conclusion, which happens to be correct, would be not only acceptable as something independently knowable, but reinforced as something also statistically likely. This then finesses determining the truth or falsehood of any of the premises... and, eventually, someone is bound to "reason backwards" to deduce the statistical likelihood of the premises from the conclusion. Indeed this latter version, in which eventually all the premises are falsified as a result of starting out with a false assumption asserted as a conclusion, is exactly what has been identified and labelled elsewhere as the aphenomenal model (Khan, Zatzman and Islam 2005)

An extreme example of the utterly specious procedure that is generated by such degenerate reasoning, and the layers of opacity that it can be used to generate, made it recently to the front page of the October 24th, 2005 editions of the *Wall Street Journal*, in an article by reporter by Jon E. Hilsenrath about how a "Novel Way to Assess School Competition Stirs Academic Row" (excerpted below). The major premise in this case goes: "public schools located in different parts of the same district produce different student outcomes". The minor premise goes: "a common geographic factor associated with the neighbourhood of schools with the best outcomes is the presence of water-streams". The conclusion-deduction is: "public schools located near water streams produce better outcomes than public schools that are not". Empirical evidence for the conclusion actually exists, and was compiled by the up-and-coming researcher-star at Harvard

University's economics department whose work stands at the centre of the aforementioned "academic row." The major premise, which posits physical location as a determining factor *irrespective of the demographics of either the student population or the teaching staff*, is either meaningless or demonstrably false. The minor premise holds syllogistic value if and only if either the major premise is non-trivial or if no other geographic factor is found to be common to schools with the best outcomes. This research actually argues that, since previous research (much of it premised on linking student or staff demographics to outcomes) produced conflicting results, an allegedly "random" correlation should be sought instead which would avoid the biases that ensured the previous research approach would produce such an indecisive wash:

> Five years ago Harvard's Caroline Hoxby, a rising star in economics, wrote a paper that reached an unusual conclusion: Cities with more streams tended to have schools with higher test scores.
> Today her work is a widely cited landmark in the fierce national debate over free-market competition in public schools. And it's at the center of a bitter dispute with another economist that is riveting social scientists across the country.
> Her adversary is Jesse Rothstein, a young professor at Princeton, who says her study is full of flaws. In a rebuttal to her critic, Dr. Hoxby wrote of his work: "Every claim is wrong." She has also accused him of ideological bias. Dr. Rothstein, in turn, says she resorts to "name-calling" and "*ad hominem* attacks" on him.
> The unusual spat has put a prominent economist in the awkward position of having to defend one of her most influential studies. Along the way, it has spotlighted the challenges economists face as they study possible solutions to one of the nation's most pressing problems: the poor performance of some public schools. Despite a vast array of statistical tools, economists have had a very hard time coming up with clear answers.
> "They're fighting over streams," marvels John Witte, a University of Wisconsin-Madison professor of political science and veteran of a brawl over school vouchers in Milwaukee in the 1990s. "It's almost to the point where you can't really determine what's going on." (Hilsenrath 2005b)

In fact, as the article more or less brings out, the real *casus belli* over this research is that the role of school vouchers for parents to choose the best school for their child is treated as a neutral and non-biasing factor, whereas the opponents of the "streams" research and its conclusions are damned by the pro-"streams" faction as an elitist coterie of anti-voucher, anti-choice fanatics. Thus indeed it turns out that the research design of the "streams"-camp has its own agenda, *viz.*, to occupy the education outcomes research space by ousting the anti-voucher elements that prevailed a long time there. At the same time, the initiator of school-voucher economics of parental choice, the authority and reputation of Milton Friedman – practically a demi-god of the Wall Street Journal famous for restoring monetary theory, dethroning Keynesian "orthodoxy" and capturing the Nobel Prize in Economics for his trouble (Friedman 1976b)– is involved, and so the reporter is compelled to turn himself into a human pretzel and couch his findings about this "academic row", which are actually quite damning for the voucher-choice camp, in terms that are as uncondemning as possible:

> Milton Friedman, the Nobel Prize-winning economist known for his free-market views, proposed 50 years ago that to improve schools, parents could be given vouchers – tickets they could spend to shop for a better education for their kids. He theorised that the resulting competition among schools would spark improvements in the system. Free-market advocates loved the idea. Teachers' unions hated it, arguing that it could drain resources from some public schools and direct resources to religious institutions.
>
> Research on these programs turns up evidence of benefits from school choice. But it hasn't proved strongly convincing, and testing the hypothesis is anything but simple. In the mid-1990s, researchers battled over how to interpret studies of voucher use in Milwaukee. In 2003, they tried to evaluate voucher experiments in New York and ended up squabbling over the right way to decide if a child was African-American. Last year, in assessing charter schools – institutions that are publicly funded but not bound by traditional rules – they argued over how to take into account differing backgrounds of the children who attend.
>
> Analysts have searched as far away as New Zealand for evidence about the effects of competition in education – and disagreed about what was found there, too. Now there is Hoxby vs. Rothstein.

Up to now, the ace-in-the-hole argument for relying on statistical procedures and processes to rigourise social science has been that, apart from investigations of extremely limited phenomena, and since results cannot be reliably duplicated where input conditions cannot be fully or faithfully replicated, lab-controlled experimental reproducibility of the kind routinely utilised in the natural sciences is really not an option in the social sciences. Does it follow from this, however, that phenomena observed in society, its politics and its economics cannot be ascribed accurately to definite causes? Instead of addressing this meat of the matter, advocates of statistical methodology as the heart and soul of rigorous social science raise their diversion that, without a probability measure, there is too much room for subjective opinion and judgment. What has either not occurred to some of them, or already been dismissed by others among them, is the idea that, instead of blithely and unquestioningly assuming that the status quo is all there is, all there has ever been and all there will ever be, such arbitrariness is precisely what could be reined in by properly and duly incorporating characteristic historical time-dependent conditions attending the emergence or disappearance of societal phenomena.

The "properly and duly" caveat is important in this connection, as it is entirely possible to arrange historical data so that one arrives at no single determinable cause or clearly-defined pathway of causation. As Gilbert and Sullivan parodied in their operetta *Trial By Jury*, the English judiciary strutting about like truly feudal nobility could assert their eternal right to control the status quo by noting that "if everybody's somebody, then no-one's anybody". In the social sciences, modern-day academic nobility – actual or aspiring – behave exactly the same when they assert that a phenomenon that they have studied to the point of practically converting it into their personal property has so many causes that no one cause or pathway can be sorted out. How do those who commit such felonies against scientific integrity and the authority of authentic knowledge get away with it? By improperly and unduly manipulating the intangible aspect of temporal factors.

4.2. Essential and Intangible Role of Temporal Factors – A Detailed Example

4.2.1. Detaching Canada's East Coast Fishery from its History: Causes and Consequences

Coming mostly from large institutions based in the United States and others around the world following their lead, there is a trend that has come to predominate in current social-science writing which generally avoids historical dynamics altogether and resorts to mining history mainly or only as a source of factual documentation of past events. The period immediately preceding that *coup* was rich with examples of exactly this kind of felonious assault upon scientific integrity, and works from that period provide some of the richest teaching material by negative example.

This section discusses one particular example at length. Of course there is much to elaborate about the particular subject matter – the east coast fisheries of Canada – which in itself is hardly a commonplace everyday subject. But the central error in its methodology is common to a very broad range of writings especially in the fields of economics, development and theories of social and economic systems and their interrelationships.

The work in question is a collection of monographs by then-young and upcoming economists, historians, political scientists and sociologists dealing with the fishery of the Atlantic provinces of Canada. Published in 1979 by "New Hogtown Press", a special imprint of the University of Toronto Press, the collection was entitled *Underdevelopment and Social Movements in Atlantic Canada* (hereafter: *U and SM*) and edited by Canadian sociology professors Robert J. Brym and R. James Sacouman (Brym and Sacouman 1979). As recently as October 2001, more than two decades after the appearance of this book, at a conference on "regional underdevelopment" convened at Saint Mary's University in Halifax, Canada. Several of the work's contributors were still actively defending the lines of interpretation advanced earlier in *U and SM*.

Veltmeyer and Petras have further developed certain aspects of the line of thought in Veltmeyer's piece in *U and SM* into a thesis concerning the purported ecological disaster of the Canadian east coast fishery as an example of what they call a "system in crisis" (Veltmeyer and Petras 2004). Veltmeyer and Petras (hereafter *V and P*) repeat the claim popularised by many observers that the shutdown of the commercial fisheries of eastern Canada by the federal government since July 1992 had to do mostly with overfishing of the resource base in the sense of causing the resource base to be reduced below a level that could sustain a similar level of fishing effort, and then they argue from this what a blight on Nature and the ecosystem such excessive plundering represents.

Is there evidence that these fish stocks were being harvested beyond their capacity to sustain such effort? Yes. Is this, however, evidence also at the same time of too many fishermen? *V and P* duck this question, by stressing the devastation of the resource base. But economics being about human livelihood, it is for our purpose an absolutely crucial question. It is not enough to consider the relation of persons to Nature when it comes to pursuing a livelihood; the relations of persons to one another in the pursuit of livelihood is equally important because, as will be discussed below, our starting point is that Nature is the mother and labour is the father of all wealth. Failure to take this properly into account has led and seems always to lead researchers away from accepting, acknowledging and reckoning the most important corollary flowing from this observation, *viz.*, that politics cannot be separated from economics and, indeed, commands it.

Behind the use of the overfishing thesis to explain or justify the government's "groundfish moratorium" of July 1992, is the doctrine repeatedly put forward by the Canadian federal Department of Fisheries and Oceans: that there were too many fishermen chasing too few fish. The scale of the event alone, however, renders that explanation immediately suspect. This was the single largest lay-off in Canadian history, in which 40,000 livelihoods were eliminated overnight almost entirely in the outports of the south, west and northeast coasts of the island of Newfoundland (as well as to a much lesser extent in parts of eastern New Brunswick, Prince Edward Island and eastern and southwest Nova Scotia). Government-collected data on fish catches inside Newfoundland and Nova Scotia territorial waters (within 12 miles of shore) disclosed a pattern of overfishing in all the principal commercial species by the early 1980s. The peak catch by Canadian fishermen on the east coast had been recorded in 1968 and this overall total had been in decline ever since (Harris 1998).

V and P mention the role of the expansion of the 200-mile offshore economic zone as a factor stimulating excessive fishing effort by Canadian as well as foreign fleets in the northwest Atlantic. However, they do not analyze or explain how the corporate sector used its connections in the Canadian government to multiply the effect of its domination of the processing and harvesting sectors of the fishery to drive the small independent fishermen to the wall, out of fishing altogether. This is significant as the contributors to the earlier *U and SM* volume had produced their papers mostly at the time, or just before, the 200-mile limit was introduced in 1977. In any event, their treatment of the general impoverishment and social-economic stagnation of the great mass of fishermen in the coastal communities over a lengthy historical period, pays no attention to the specific impact of the vertically-integrated corporate sector on how so many coastal fishermen continued to harvest most species. They manage this independently of any relationship with these processors, certainly at increased cost to themselves and hence also at the cost of continued marginalization of their incomes from fishing. None of the papers reflect any consciousness of its imminence even though the Canadian government had made clear as early as 1975 that it would proceed with such a policy once it negotiated and signed bilateral fishing agreements with the largest of the 18 foreign fishing fleets, led at that time by the Soviet Union, harvesting catches within 200 miles of the Atlantic coastline. The bilateral agreement reached between Canada and the Soviet Union in 1976, three years before the appearance of *U and SM*, actually provided its fleets could take up to 88.4 per cent of the fish stocks inside the 200-mile limit on the east coast, according to the principle that Canada treat all fish stocks not being commercially harvested or developed at the time of the agreement as "surplus to Canadian needs". Such vending of offshore sovereignty registers nowhere in *U and SM*, a circumstance that seriously blighted any possible value of their analysis. At the same time, this is hardly surprising as the body of economic theory underpinning that work nowhere acknowledged even the possibility of rival imperial interests – in this case, the U.S., and then-Soviet Union – contending for supremacy and privileges in one and the same zone, *e.g.*, Canada's northwest Atlantic littoral, lying beyond the formal territorial control of either Washington or Moscow.

The table of contents of *U and SM* discloses its general drift:

1. The Capitalist Underdevelopment of Atlantic Canada, by Henry Veltmeyer;
2. The Differing Origins, Organisation and Impact of Maritime and Prairie Cooperative Movements to 1940, by R James Sacouman;
3. Political Conservatism in Atlantic Canada, by Robert J Brym;
4. The Emergence of the Socialist Movement in the Maritimes 1899-1916, by David Frank and Nolan Reilly;
5. Underdevelopment and the Structural Origins of Antigonish Movement Cooperatives in Eastern Nova Scotia, by R James Sacouman;
6. Underdevelopment and Social Movements in the Nova Scotia Fishing Iindustry to 1938, by L Gene Barrett;
7. Inshore Fishermen, Unionisation and the Struggle against Underdevelopment Today, by Rick Williams;
8. The Capitalist Underdevelopment of Nineteenth-Century Newfoundland, by Steven Antler;
9. Regional Factors in the Formation of the Fishermen's Protective Union of Newfoundland, by Robert J Brym and Barbara Neis; and
10. Towards a Critical Analysis of Neo-Nationalism in Newfoundland, by James Overton.

In addition to the broad topic of Canada's east coast fishery, these essays share two other links in common. First, they share an acknowledgement of the need to incorporate historical reality. Second, but unfortunately coupled with the first, they also each fail to think through the impact such a procedure must have on the manner in which the more conventional tools of their particular discipline are handled and applied. As this failure forms the starting point of the discussion rather than its conclusion, it is appropriate to set out the starting point adopted for the purposes of the present critique.

As a social political and economic fragment, observed from the vantage point of the present, Canada's east-coast fishery can neither be separated from national and international politics nor reduced to a riot of spontaneous parochial conflicts. Looking back into the past from the vantage-point of the present, it can be seen that, from its inception during the dawn of colonial expansion into the northwest

Atlantic approaches of the North American continent at the end of the fifteenth century by Europeans, politics has always commanded economics in Canada's east-coast fisheries.

The "conventional" approach taken by government and industry economists, however, assumes the vantage-point of the present mainly in order to focus on the future understood in the most immediate short-term. As a result, this approach tends to separate the politics of the fishery from economics. Struggles, contradictions and other disturbances are viewed as aberrations from the norm, rather than as natural products of the politics and economics of the fishery. The contributors to *U and SM* write critically about the politics, economics and history of the east coast fishery – and at this point in their careers mostly independently of the policy-setting apparatuses of government or industry – taking a "critical" approach. They concentrate on the struggles, contradictions and disturbances of this sector, but narrow their focus to the immediate, current, contemporary social and economic conditions giving rise to such struggles.

Here, however, in the narrowness of this focus lies a most serious problem. Such an apparently analytical and even "critical" approach loses sight of the big picture in which politics and economics are acting in combination. In fact, it reduces the fishery to a riot of spontaneous parochial conflicts, in the sense that these points mark the contours of its development path. Wittingly, or unwittingly, however, such a procedure incidentally also jettisons information – historical data – about both the persisting continuities from the past into the present as well as about whatever phenomena discontinued during the passage from the past into the present. By making so many pertinent facts disappear, this jettisoning has created large gaps in understanding and provoked researchers to suggest dubious explanations and conclusions.

Consider the fishery in the present-day provinces of New Brunswick, Nova Scotia, Prince Edward Island and Newfoundland and Labrador, taking one snapshot in the last third of the 19th century and another in the last third of the 20th. In the late 19th century, nearly all labour in these fisheries was part-time and-or seasonal, the vast majority of it unwaged. The workforce of than 1.2 million persons comprised more than 80 per cent of the total population of these provinces/colonies. In the late 20th century: all labour, full time or part time, had become waged. The total population living off the fishery was cut in half. The fishery labour force itself had been reduced to one-tenth of the population resident in the coastal fishing communities. The value of primary and processed output had increased more than 100 times. Some 90 per cent of that value was being produced by about one-fourteenth of the entire labour force directly employed in the fishery. The rate of profit generated by the exploitation placed on the shoulders of this fourteenth part had become extremely high compared to the rate and level of profit generated from the same level and intensity of exploitation applied in various other ways to the other 13/14-ths of the labouring population. Over this entire period, regardless of the dramatic transformation in productivity represented by these data, the standard of living and working conditions for more than half the population of the fishing communities continued to fall behind the Canadian standard.

All these differences in income and huge differential in rates of exploitation between the fishery and sectors of other economic activity in Canada are symptomatic of an economic order in which overall development is characteristically highly uneven. This unevenness reflects both the degree to which, as well as the manner in which, material production and its ownership have become intensively concentrated in fewer and fewer hands. The same can be said of the large disparities in income and living conditions that persist and continue into the 21st century to grow both between the best-off parts of the fishery and the worst, as well as between the best-off parts of the fishery and the Canadian standard.

For all this transformation, much of it quite marked, certain invariant features persisted to this day. What remained more or less the same was the proportion of Canadian-generated output from these fisheries that went to markets outside the country. Furthermore, the fish harvesting effort initiated from coastal communities in these provinces/colonies in the northwest Atlantic fisheries always remained less than the effort mounted by the foreign fleets catching fish in these same waters. These invariants are not discussed or mentioned by any of the contributors to *U and SM*. Something somehow inaccessible to their methods of evidence-gathering or detection seems to be going on. How else can one account for these invariants on the one hand and their non-observance on the other? This was a fishery whose very discovery by European capital in the 15th century contributed to creating a world market in foodstuffs in the first place and whose existence and operation continued on the basis of participating first and foremost in this same world market even more than 100 years after its occupation by Canadian fishery enterprises and workers. Apart from a stale reference to (and dismissal of) the work of the Canadian social scientist who pioneered some initial investigation of this global aspect in the 1920s (Innis 1954), in the work of *U and SM*'s contributors there is no consciousness whatever reflected anywhere of how highly remarkable such an economic fragment is – either for the part of Canada in which it operates, or for the international standing and role as a whole of Canada itself. The fact that some sea-change took place since that time is obvious only from the vantage-point of the present looking back. On the other hand, any attempt to account for either the fact or scale of this evidently intangible transformation by methods that rely upon or use t_{LINEAR} – a linearised conception of time moving forward through some interval starting at $t_{initial}$ and ending at t_{final} – would completely miss this transformation.

What constitutes historical perspective and how can it be incorporated as objective historical data in social science research? There are various ways to look back into history. One might look back on the basis of meeting what is found in the past "organically", so to speak, on its own terms, *i.e.*, suspending one's own contemporary understanding of what was unknown, misunderstood or not recognised back in the period or at the time of interest. Alternatively, one might look back in a linearised way, taking the present as t_{final}. Historians and economists in the main have been systematically trained to follow, almost unconsciously, the latter path. The mindset and assumptions that accompany the conventional presentations prove it: the historical orthodoxy is that this fishery itself arose spontaneously as an accident of European discovery, while the economists' orthodoxy is that its commerce and industry developed as an epiphenomenon of Adam Smith's "invisible hand".

Taking the "naturalised" historical viewpoint, however, a different picture emerges. From the end of the 15th century to the middle of the 18th, this activity emerged as the conscious, not particularly well-planned, and highly contradictory outgrowth of competing Great Power schemes of colonisation. There was the global Roman Catholic missionary agenda of the "united crown" of Ferdinand and Isabella of Spain. There was the struggle of the English crown to free its merchants' activities in the "Western ocean", *i.e.*, the Atlantic, from the control or interference of the Spanish navy. There was also the struggle of the French monarchy to colonise the "New World" as a means of bolstering its absolute rule over an increasingly fractious feudal nobility.

As a result of the aforementioned information loss generated by the so-called "critical" approach, however, and irrespective moreover of the differences between the approach-path of the "critical" analysts and that of the "policy wonks" in government and industry, the window in which to observe and note the big picture as it actually plays out becomes shattered into countless fragments. In that shattering, politics once again becomes separated from economics. In that separation, yet more information is also lost about the meanings and intentions of the various actors at different times and places in the east coast fishery. Some of this editing is deliberate, based in a confusion about the supposed need to restrain or prevent excessively subjective modes of interpretation being imposed on the "bare facts" of "history". Obviously, however, at the time these actors acted, history was neither yet history nor "bare facts" but a set of problems being taken up for solution.

4.2.2. Mishandling Temporal Factors: A Problem of Method

The actual development of Canada's east coast fishery cannot be reconstructed without analysing and re-synthesising, **in real time**, the interaction and consequences of the human labour of fishing on the social as well as the natural environment which surrounded it and which supplied and enabled this activity in the first place. "Real time" means not necessarily the present, but as things actually happened: starting with the origins in its actual history and peculiar conditions. The unfolding of the key moments of human history surrounding its development must be faithfully reflected. Thus, for example, in the beginning there was the colonial expansion and ambition of states, as public entities, to increase their power relative to rivals and competitors. There were the ambitions of various individuals and groups (in a capacity or station distinct from the state) to acquire enormous private wealth by joining the trend. This combination encountered the Grand Banks fisheries off Newfoundland at a time of Great Power contention over who would control the riches of the New World (which included maintaining a vast international traffic in slave labourers). Our main concern, however, in the present work – which is neither the time nor the place for accomplishing such reconstruction – is to illuminate the failure of the *U and SM* contributors' analysis of the transformation of the socio-economic organisation of the Canadian east coast fisheries **in order to disclose how that failure is linked to mishandling intangible temporal factors**. This mishandling leads to glossing over, mis-stating, misinterpreting or missing altogether the causes and consequences of these transformations and thereby holds out larger lessons and warnings for contemporary social science research in general and as a whole. This task is essential for clearing the path to elaborate on a scientific basis the promise of the present volume, *viz.*, an "economics of intangibles".

Consider what would be involved if one were to re-till the ground ploughed up by the contributors to *U and SM* with the aim of re-doing the work. In order to eliminate misleading, unwarranted or demonstrably false conclusions and inferences developed in the essays of the present work, two tasks would urgently present themselves:

1. to review the *U and SM* contributors' selection of dynamically important criteria that influenced the actual development of the conditions of the

past towards what emerged as the conditions in the present, re-examining them from the standpoint of determining their correctness or incorrectness relative to how faithfully and especially how non-anachronistically their use of historical materials and dynamics actually reconstructs the past; and
2. to review the contributors' selection of dynamically important criteria influencing development in the present whose origins are to be specifically located in the past.

The difficulty is that these errors and sources of error are bound up with the banner of "radical critique" which the editors of the *U and SM* project planted throughout. If one were trying to re-do this work on a consistent and scientifically sounder basis, it would not do to become become bogged down in refutation of details. Nor, however, would it do to simply wave an opposing flag. The issue here would become: how to show one's colours through the deed of taking a clear-cut stand restoring scientific integrity in social science. In reference to this specific work and its scope, it is here we encounter the nub of the problem, *viz.*, how does one tackle, *i.e.*, break down, the notion that "critical" = "Marxist"? That is, setting aside any urge either to purify the Marxism of the contributions to *U and SM* or eliminate it, how does one overcome the syndrome according to which donning the mantle of "critic" also confers a licence to recycle assumptions in the name of scientific method or of "Marx's method" – assumptions and methods that, upon further scrutiny, turn out to be indistinguishable from the assumptions and methods of those who were being attacked for the conventionality of their approach or the narrowness of their service to the interests of industry and the state?

There are two issues involved with these "critical" essays. On the one hand, as far as existing approaches are concerned that purport to explain the status-quo by affirming it, critique has a positive role to play. On the other hand, no amount of wielding of the categories of some method (in this case, Marx's actual method) at one's opponents is going to penetrate social reality to its roots and faithfully represent its actual processes of change, development and motion in their all-sided profundity. Given our concern here with what happens to scientific method in general and the optimal use of historical dynamics along with historical facts as temporal factors in particular, this is a matter of some moment. What is needed is both time as it is experienced in living reality, and time conceived historically, *i.e.*, over periods that may exceed many lifetimes. The former without the latter eliminates all perspective. Going down that road, we may as well all proceed to join the Flat Earth Society as to conduct serious further research in economics or any other field of social science. The latter without the former, on the other hand, renders the experience of economic reality inaccessible and unreal and disconnects the long-term from the short-term or "immediate reality." That leads to "science" that is useless. The only path on which scientific integrity can be maintained for this task is to revisit some fundamental definitions.

Throughout scholarly discussions of economics, the conventional metric adopted for the concept of wealth and its quantification actually marginalises the role of Nature in the extreme. Neither the conventional nor the avowedly "critical" approaches to Canada's east coast fisheries, for example, uphold the principle that Nature is the mother and labour the father of wealth. This rejection is implicit in their mode of presenting value-in-exchange of fishery products in the marketplace as the prime concern at the business, commercial end of the fishery which drives all other concerns and thereby leapfrogs having to examine the value-in-use both of fishing as an activity as well as of fish as food. The idea itself, of the complementarity of Nature and Labour, other versions of which appear in ancient Greek philosophy, only came to be formulated this way in the late 17th century by the English writer William Petty (1678). Yet it encapsulates a truly time-tested principle – *viz.*, that wealth itself is something not to be hoarded but first and foremost to be "created", *i.e.*, fashioned, from raw materials worked upon by human labour. This principle takes into consideration that economic activity which produces what can properly be considered "wealth" comprises relationships not only between between people and Nature, but between a person or persons and another person or persons. What, however, happens the moment both the conventional and critical factions of the fisheries discussion shunt this old-fashioned idea aside? This is precisely the point at which politics gets separated from economics, with Nature (in the form of waters and the fish) and Labour (the fishermen) banished to the periphery. Although the "critics" dispute many of the conclusions of the conventional economists, they never challenge the conventional economists' basic method. As a result of the fact that they share a common approach with the conventional economists, the critics in every one of their contributions to *U and SM* end up conciliating the separation of politics from economics. The problem with such separation is that, once politics – the matter of interests and especially of intentions – is removed, economics is reduced to considerations of time t = "right now".

Conventional economic science remains insistent that such separation ensures an economics that is stands above, and remains untainted by, the crass conflict of competing political interests and intentions. Is this now, or has this ever been, the case? On the contrary: it actually politicises scientific inquiry in the worst possible way, by weaving together scenarios and explanations that enshrine the TINA syndrome. As economic existence – the winning of mankind's bread and livelihood from participation in social labour – is an arena in which constant turmoil, and not the steady state, is the norm, the absurdity of TINA-type analysis and conclusions based thereon is self-evident. At the heart of this lofty pose of objectivity and standing above the political fray there is nothing but a rabidly fanatical ideological commitment to the status quo, no matter how much "science" is mustered in justification. The approach taken by conventional economists to the fisheries universe is the same as the approach taken by Ptolemy and the mediaeval Vatican to the physical universe. According to Ptolemy, using crude instruments and guesswork, the Earth was at the centre of the universe and the sun and the rest of the heavens revolved around it. With better instruments and more precise guesswork, Ptolemy might have junked his erroneous initial guess and hypothesised otherwise. We will never know for certain, but at least he advanced his hypothesis on the basis of what he thought were sound and verifiable observations. According to the Vatican, however, everyone had to accept Ptolemy's conclusions without question — regardless of the findings of science and observation to the contrary centuries later. This approach to matters of science can be faulted on two counts. First, a key assertion is accepted as fundamental without further testing. Second, the assertion of a preference is permitted, encouraged and upheld regardless of the evidence of objective, material reality. The starting point of serious scientific enquiry, however, cannot be the wishes of any individual or group, however just or unjust. Objective phenomena have material causes and effects that have to be observed and accounted for as they actually are, not as anyone might wish them to be.

Not interested to deal with actual historical development, however, proponents of both the conventional and "critical" approaches choose instead to place the enterprise and initiative of individuals or corporations at the centre of their fisheries universe. In effect, this is an implicit declaration that whatever is true for individuals counts for more than whatever the truth of the overall picture discloses. This removes the problem from the frying pan, however, only to toss it onto the open fire. As a scientific matter, the political economy of the fishery has to be explained in terms of the relations of cause and effect as they develop in the actual material conditions. These conditions are something in which the will of individuals or companies plays some role, but only in the context of the objective laws of motion guiding that system in a certain direction at a given time, regardless of what those exercising or chronicling that role may choose to believe. This cannot be an overall determining role. Even if one starts from enterprise structures and-or functions, this must be done non-anachronistically. It must be done in a manner that that will remain faithful to how the reality unfolded in historical time. Developments must be traced from their emergence on the margins of the European commercial and slave systems to the present day, out of conditions in which the merchant was the factor outfitting and equipping the producers who organised all the actual production functions, to conditions where vertically-integrated units of finance capital emerged in dominant roles.

Men make their own history: for the conventional approach this is enough. A confusion arises on this point, however, when it comes to the work of proponents of the "critical" approach. Karl Marx himself issued a famous caveat about men making their own history, *viz.*, that this takes place in circumstances already shaped by actions of others in the past and thus not in circumstances entirely the choosing of those acting in the present (Marx 1859). What is missing from the contributions to *U and SM* is any consciousness about what it means or how to apply this criterion to the material in question. When it comes to making due and proper use of intangible temporal factors, what becomes crucial is the implication of Marx's caveat, *viz.*, that material systems of relations of cause and effect have their own laws whose structure as a system then shapes how contending interests form their will as well as how they may implement that will.

In the process marginalising the role of both Nature and Labour, the conventional economists' approach seeks categories of discourse that are as bloodless as possible. Instead of explicitly differentiating causes from effects, they revert to eclecticism. Frequently some effect is blamed on a multiplicity of causes of different kinds and qualities. If no single cause can be found to account for all facets of a phenomenon, then — according to this line of reasoning — there cannot even be some single cause that would account for just the principal features or essence of the phenomenon. Here is the point at which Science is grabbed by the lapels, beaten up and hurled into a dark alley to be left for dead as its positions are simply usurped... by Solipsism. Instead of zeroing in on intentions – on correlating the negative consequences with ill intentions and positive consequences with good intentions – matters are reduced to the supreme Solomon-like judgment of the omniscient individual. The individual as ultimate arbiter in charge of assigning Causes and Effects is a scenario that has put in its appearance in many a "study" of the problems of the Canadian east coast fisheries, but here one example will suffice. In 1982, the Kirby task force on the Atlantic fisheries produced a report entitled *Navigating Troubled Waters* which declared that, when it came to differentiating actual causes from effects in the real world fishery, "where you stand depends on where you sit." (Kirby 1982) In other words, everyone had an axe to grind or special interest which would colour their analysis. That is likely true, but the unwarranted further conclusion extracted from this is that nothing could be sorted out objectively. This sets up a scenario in which Canadian fisheries expertise, backed by the government, plays Solomon in resolving the contradictory claims of livelihood from the fishery in the coastal fishing communities on the one hand, and profits from the processing and sale of fish products in the boardrooms of the corporations involved on the other. This unwarranted conclusion is based on assuming precisely what was yet to be proven or disproven after an objective weighing of the evidence. Such an approach yields a variant of what has been identified elsewhere as the aphenomenal model (Khan, Zatzman and Islam 2005). It ensuries that no problem will be analysed to its root and solved by sorting out actual causes and effects.

The situation with the economic theory and analysis of the "critics" contributing to *U and SM*, however, is still more complicated. Explicitly they affirm and start by placing the situation facing the people at the centre. They do not hesitate to criticise openly the conventional wisdom that consigns these concerns to the periphery. Appearances, however, are deceptive. The problem starts when they posit the situation in the fishery in terms of something they call "regional (capitalist) underdevelopment" (hereafter: RCU), blaming everything that doesn't fit the norm of the conventional experts on, or ascribing it to, this "underdevelopment." As will be shown, this recapitulates the error of conventional economists' eclecticism. Once again, it blames something on "everything" and thereby on nothing, while at the same time also denying any role of intention, and it does this no less systematically than the conventional economists. Its implicit utopian and unwarranted assumption is that an ideal world would provide full economic planning including some rational restraints on freedom of movement for Capital. Although RCU is put forward as a reinterpretation and refocusing of the historical background, and the present and future of the east coast fishery, it is not advanced on the basis of any actual or thoroughgoing deconstruction-and-reconstruction of that history. Rather it is based on a much more limited approach of arranging and rearranging key developments. Notwithstanding the burden of responsibility and quasi-magical powers with which these critics have invested RCU, however, this makes these essays just as instrumentalist as any conventional economist insisting on working within an unmodified status quo. No less than any conventional economist, these critical essays are just as unconcerned to get to the bottom of matters, to establish and distinguish actual causes from their effects. Not surprisingly, therefore, do they end up eventually affirming the TINA syndrome albeit in modified form.

2.3.2.3. Social Science and the Problem of Linearised Time

All this poses the question: how could such loudly proclaimed progressive social commitment end up affirming the status quo? Here there is plenty for historians of political ideology to mine. In the present work, a little further on, the historical context in which RCU theory emerged is set out. Of more immediate concern, however, is the evident problem such a state of affairs suggests regarding the methods of these "critics" as social scientists. The immanent cause of the difficulties and contradictions rending *U and SM* is a notion that one becomes the most radical critic of everything existing merely by donning the mask of The Most Radical Critic Of Everything Existing. However, neither dismissing this caricature of "Marxism" for being a caricature, nor pointing this out and condemning its immaturity, gets to the bottom of matters. That there is more than enough Marxism in these critics' madness cannot overcome serious deficiencies in their method. The essence of the problem that will Inow be examined lies with that method. In its attempts to attack and expose ongoing effects of European colonial expansion in newly independent countries, this critique has borrowed some terminology from Marx, but otherwise it has nothing to do with Marxism. Most tellingly of all in the context of the present work, this critique fails to identify the core of the anti-Nature, Eurocentric outlook responsible for the full-scale assault unleashed by that process of expansion against, and at the expense of, Humanity's prospects. For these critics, the times are literally out of joint. What is required here is to tackle precisely whatever is responsible for the improper and undue manipulation of the intangible aspect of temporal factors on display in this collection of essays, and use such deconstruction to construct or point to a proper method.

To accomplish this entails:

1. an examination of the intangible aspects of temporal factors involved in the introduction and expansion of the colonial system to the "New World", *i.e.*, the Americas, which serve to particularise the emergence of full-blown industrial capitalism as a world system while at the same time differentiating its actual development in specific sectors and regions;
2. an examination of how and where, in the context of the contradictions that emerge in Canada's east coast fishery, the intangible aspects of temporal factors operate to universalise some elements of the operation over the passage of time of the industrial capitalist system as a whole, but not others; and
3. an uncovering of the source of much mischief-making lying at the heart of efforts to rationalise and-or justify the TINA syndrome, in the massive confusion surrounding the intangible aspects of temporal factors and their significance in general. This includes in particular the significance of the

differences of t_{LINEAR} in all its various forms from all forms of $t_{NATURAL}$, including $t_{HISTORICAL}$ which is the form that $t_{NATURAL}$ takes in social science.

Establishing points (1) and (2) is straightforward. A limited amount of historical exegesis is sufficient to compel a broadening of consideration, even a partial reconsideration, of the manner in which the expansion of the North Atlantic fishery played its role in the expansion of European capitalism during the early modern period. The third point, on the other hand, raises directly the question of the context and applicability of t_{LINEAR} to historical phenomena. This is indeed a problem that is profound and complex at the same time. Later this analysis will go into some detail as to how far this distorts the entire fisheries problematic. At this point, it is possible and perhaps necessary first to introduce what is at stake as far as scientific method is concerned.

Consider the category "accumulation of Capital". Whether as a mass of exchangeable value or as a collection of exchangeable values, accumulation of Capital may be investigated and summarised objectively (in the sense of "independently of anyone's will") as a function (or set of functions) of some independent variable that will assume one or another form of t_{LINEAR}. By its very nature, the important things to know when attempting to measure accumulation of Capital are the starting and ending points of whatever the selected time interval. It would be entirely expected of anyone investigating this scientifically to wield the tools of Newtonian calculus, utilising an independent temporal variable in t_{LINEAR} form.

Within this, however, there lies an interesting, and remarkably unremarked, paradox. In capitalist societies, there is only the aim of maximising individual wealth and no overall *societal* aim. Overall economic development proceeds through cycles of time in which there are innumerable branch-points. Such an inherently non-uniform time span cannot be classed as a form of t_{LINEAR}. Nevertheless it is considered entirely reasonable to examine certain specific epiphenomena within such cycles, such as "accumulation of Capital", still utilising one or another form of t_{LINEAR}. This approach may even be extended to define the total capital of such a society as the sum of all the individual capitals thus accumulated.

When it comes, however, to comprehending *changes of state*, as it were, what justification remains for continuing to rely on any form of t_{LINEAR}? (By "changes of state, we have in mind" what takes place, for example, in the overall cycle of investment boom, overproduction and crash, or in the movement from one cycle to the next, or in the emergence of innumerable unintended consequences, especially the rising impoverishment and related degradation of the general population that arise during and as part of these cycles.) What basis is there to assume a similar applicability for a linear scale of time in which only the starting-point, ending point and duration in between are of interest? While the recurrence of the cycle ensures that these phenomena will also recur, the mere fact of such recurrence in itself neither predicts the onset of these manifestations of the anarchy of production, nor the exact position in the cycle of the branch-point associated with such onset (Kondratieff 1935).

This pattern for the category "accumulation of Capital", which is a category that is central to any and every part of the capitalist system, can be repeated for any other individual category. This suggests that t_{LINEAR} works adequately when it comes to looking at rates of change for any element of individual capital. However, it cannot provide a significant source of non-trivial information about the larger societal picture. Under modern capitalism, how do prominent individuals continue to play leading roles? They do so no longer in and of themselves, or – speaking more objectively – in or according to the amount of capital they personally represent, but rather as agents of a grouped, collectivised corporate capital, a capital that has been assembled by expropriating large dollops of social capital through government connections, membership in interlocked directorates of corporate and bank boards (Mills 1956). The relatively more prominent role traditionally assigned, from an earlier stage of capitalist society, to the individual (over and as opposed to social collectives) cannot account, however, for this difference in the viability of basing serious analysis on forms of t_{LINEAR}. Applying the principle of "Occam's Razor" – the principle of reasoning according to which conclusions which would follow from the available evidence (as opposed to wherever anyone might wish the evidence to point) provide at least a first approximation of the truth – we find a more compelling reason. No form of t_{LINEAR} can provide useful information about developments or categories that are functions of such collectives.

Above all, however, the question "why accumulate?" is not posed, nor are the intentions of those who would accumulate explored. The essays in *U and SM* typify an opposing standpoint – that the temporal metric, linearised or historical, is a matter of indifference. The arguments of its contributing authors are all based on the assumption that political-economic *function* is entirely a matter of socio-economic *structure*, with time simply passing along like some classic Newtonian independent variable. According to this viewpoint, it would make no difference whatsoever, analytically speaking, if "the price of widgets in Slovenia" were substituted for historical periods incorporating the profoundest social transitions. The question of intention being neither asked nor answered, there is no way to satisfy any of the fundamental demands of the polity for accountability in the sense of the taking of social responsibility by individuals for the consequences of their actions in the fisheries sector.

The "accumulation of capital" category itself provides no information whatsoever about the actual development of any of the new resources generated by and available to a society in collective forms as the result of expanded investment in material production as such, *i.e.*, in the production of the use-values humans needed for societies to sustain themselves as human societies. Such socialisation of produced wealth completely transforms the temporal metrics needed for measuring progress and detecting leading or lagging indicators of where the society as a whole is headed. In particular, t_{LINEAR} becomes useless and meaningless as the length of the cycle required to reproduce and replace needs is itself shortened many times over and transformed by the new forms of social organisation made possible as the result of such collectivisation of the entire social product. The secret of this difference – one which is seen everywhere these days in Cuba, for example, and widely recognised and commented outside that country – lies in the conscious decision that is taken whether to prioritise on the one hand the accumulation of wealth in the form of exchangeable value, either (in the case of developing countries) ahead of all other considerations or (in so-called more "developed" economies) to the exclusion of all other considerations, or whether to prioritise instead the capture and achievement of the intangible social, long-term benefits lurking potentially within any socially-organised form of material production.

The development of objective descriptions of relationships in social science was profoundly affected by the fact that t_{LINEAR} *à la* Newton had been monopolising European scientific discourse from the early 18th century onwards. Even $t_{NATURAL}$ was partially fitted by resorting to periodically predictable regularly-spaced cycles, while exponential time was readily fitted by means of Euler's famous discovery that $e^{i\pi} = -1$. Other timescales or models of time were adapted to fit these parameters. Those that did not or could not fit, like $t_{HISTORICAL}$, were by and large dismissed. Such a marginalisation of reference-frame scrapped a potentially huge source of information of a kind obtainable in no other form.

This loss is not a purely passive one. Marginalising the reference-frame is also a tremendous weapon to wield against the challenge that a new discovery might pose to established knowledge. The struggles waged in European intellectual circles throughout the 17th, 18th and 19th centuries may no longer have involved stakes as high as they had been during the Catholic inquisition of previous centuries, but the struggle to establish scientific method and differentiate scientific investigation from self-interested assertions by persons said to speak with "authority" was no less intense just because it now stood at a certain remove from life-and-death. Instead of the immortal soul of the individual, what was now at stake was the sovereign claims of a social order based on private property and the right, and especially the untrammeled freedom, of those possessing private property to exploit those lacking in private property. There was now to be no freedom higher than this freedom, and this included freedom to research and establish the truth. Here was laid the foundation of all subsequent aphenomenal modeling in the social sciences (Khan, Zatzman and Islam: *ibid*).

Accordingly, by the middle of the 19th century, the challenges posed to established notions in particular by the works of Karl Marx and Charles Darwin were not small. Darwin's explanation of speciation was particularly subversive. The emergence of new species only made sense as the non-linear outcome of a lengthy series of processes that must precede and prepare the way for the emergence of a new species. At the same time, knowledge about these earlier processes, no matter how complete, still would not enable a specific and absolutely reliable prediction of all the features expressed in the new species. The story of how upsetting this was to a few religious figures concerned about the authority of the Biblical story of Creation is an old and well-told one (Irvine 1955). The upset actually went much further, however. One of Darwin's closest collaborators was the geologist Sir Charles Lyell. For the first 10 years after Darwin published his landmark work, Lyell would not publicly defend the theory of evolution. Fear of unknown consequences outweighed any other consideration, including even the fact that Lyell's own work established the notions of the fossil record and geological time the fact that Lyell encouraged Darwin through the more than two decades that would elapse between the completion of the voyages of the *Beagle* to the Galapagos and readying his *Origin of Species* for publication. Until he openly defended his friend, he officially retained public doubts about Darwin's assertion of the mechanism of "natural selection", even as Darwin was corresponding with him about these ideas (Darwin 1892).

Marx's approach to social science was even more problematic: the New would come out of the struggle between contradictory tendencies in the situation that preceded its emergence. However, the necessary precondition for such a struggle to develop in the first place was the entire historical development preceding the outbreak of that struggle. Thus the New must inevitably include some aspects of the Old, while the struggle to get there jettisoned other aspects of the Old (Marx 1859). There was one and the same message, being delivered from two very different fields. It was a message fundamentally challenging the very foundations of t_{LINEAR}: the New, or the Future, far from being mainly or only an incremental superposition on the past or the present, is a quantum break away from both.

In the natural sciences during the 20th century, work continued in many fields using t_{LINEAR} à la Newton. Some theoretical work on the frontiers such as Einstein's theory of relativity seriously tackled, at the level of the universe, the need to correct, at least in part, Newton's assumptions and implications about temporal factors and to render time's irreversibility explicit. Other theoretical and applied work such as quantum mechanics took the path of applying probability measures of uncertainty to the coordinates of elemental matter at the inter-atomic and sub-atomic levels. In general, the response in the natural and engineering sciences to this exposure of the inadequacy of existing temporal reference frames was neither uniform nor coherent.

In social sciences, there were also consequences, but the script ran somewhat differently. By the end of the First World War, the urgency of the scientific challenge represented by Marx's work as a student of economics and history was greatly increased by the emergence of actual political revolutionary movements. These challengied the established order based on private ownership of the means of production and even overthrowing longstanding regimes such as that of Tsarist Russia, the country that had become the bulwark of world reaction following the defeat of Napoleon and the settlement of European diplomatic arrangements at the Congress of Vienna in 1815. Anything considered serviceable to any part of the revolutionary agenda was thereafter branded either "communist" or outside the proper sphere of concern of economists (Böhm-Bawerk 1898). Attempts thereafter to update Marx's analysis of capitalism in conditions of free competition in order to account for changes introduced as a result of the subsequent suppression of free competition and its replacement by oligopolistic and monopoly-like "competition" waged an uphill and indecisive battle for academic acceptance (Hilferding 1910). In contrast to the general incoherence spread throughout natural and engineering science, this resulted in what might best be described as a reactionary coherence.

This reactionary coherence, meanwhile, did little for the reputation or image of work in these fields as science. Since the Great Depression of the 1930s, an endless volume of policy-related number-crunching and bean-counting took place in these fields. However, during the 1960s and 1970s, rebellion against this condition came into full flower in universities across the Americas and Europe. A parallel condition emerged, during the Cold War just before this rebellion, with the rise to power of the Khrushchev group in Moscow and their acolytes in the member-states of the Council for Mutual Economic Assistance (COMECON). Among social science academics in Soviet bloc countries in this period, as well as in developing countries sympathetic to the Soviet side, the reduction of western social science to policy-related number-crunching and bean-counting was reproduced as policy-related number-crunching and bean-counting for "market socialism" schemes along the lines opened up earlier by Oskar Lange (1938). This was very much promoted as an advance out of the alleged straightjacket of official Soviet ideology (Stalin 1952).

The result was a "renovation" and revival of Marxian thinking in social science in the West – but on the basis of t_{LINEAR}–based models. A new common ground was discovered for "Western" post-Soviet and eastern-bloc post-Stalin Marxian social science researchers. Henceforth, they would both proceed from the idea that socio-economic structure discloses everything needed to understand political-economic function. In effect, the lifetime of t_{LINEAR} would be extended in order to rationalise these changes of direction in the socialist camp with theories about a "third way." This "third way", still alive in our own time as the favourite doctrine of "new Labour" under governments led by Tony Blair in Britain, an economic course was to be plotted and followed outside capitalism or socialism, one that might be open as well to former colonies and semi-colonies of the Great Powers. Such was the setting in which theories about "underdevelopment" emerged (Baran 1957).

4.2.4. The Dialectic of Nature and its Usefulness for the Social Sciences

Nature is a dynamic environment in which changes are continual, but as a result of a strong Establishment bias against looking at processes by taking change as primary and stasis as exceptional, models that have attempted to account for the process of change have been ignored and dismissed in favour of models that attempt to account instead for tendencies towards equilibrium and the steady state. Even if claims for the primacy of change are acknowledged over those for the steady state, however, the continual changes found in natural processes are not necessarily continuous in the specialised mathematical sense. Smoothwise continuity in any natural process may be an appearance, *i.e.*, an illusion, or it may provide an approximation of a certain limited usefulness, but it seems highly unlikely as an accurate or useful description of what is taking place inside or within any moderately complex natural process overall. Relative to the particular phenomenon or phenomena under observation, Nature is observed generally from a stationary or quasi-stationary position "outside". Various hypotheses can be tested to account for the pathways that took a process or phenomenon from input to output, or from starting point to end point, but – because of this often largely unbridgeable barrier of outsidedness – complete information in many cases is unlikely to be attainable from direct observation alone.

Even when the modeler has the integrity and humility to admit that this in itself is in fact only approximating even observable phenomena, and not describing them fully or precisely, continuity-based mathematical modeling can still mislead to the extent that it sustains an unwarranted reassurance that it is of little or no moment whether the point of change within any natural phenomenon is more often like a non-linear switch or cusp-point rather than monotonically increasing or decreasing in some smoothwise continuous way. Clearly, however, if the observed pathway of a phenomenon actually looks more like

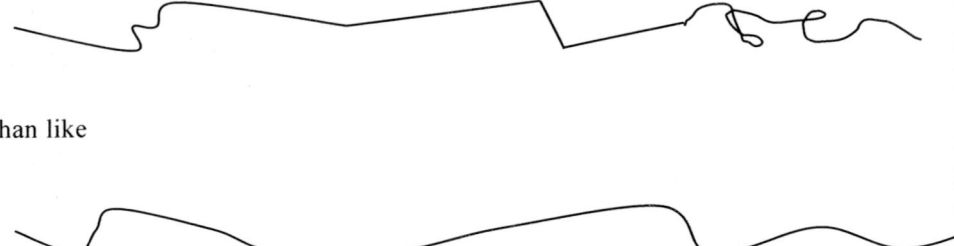

than like

it would seem a safe bet that whatever is causing the dips and changes in the first pattern is not what has given or could give rise to the second pattern. At the same time, it is also evident that the first pattern could be stretched and smoothed to produce something akin to the second, which could subsequently prove entirely misleading – especially if such mathematical tractability became installed as a criterion for deciding between equally aphenomenal "explanations" of the process being modelled.

On the other hand, if this drive to linearise is replaced from the outset by a recognition that change within a process is continual, then the basis, or cause, of change has to be clearly differentiated from the conditions in which the particular change is enabled, blocked or cancelled out. This is hard work; numerous trials may be necessary, and all this before there is any mathematical model in sight! These conditions are never internal to a process, however, and the basis of a change is never external. The conditions that enable a change to go through are external, but what could impel change in the first place is not. Thus, any conceivably applicable mathematical model is going to be inherently non-linear and must be capable of generating some discrete a number of degrees of freedom, guaranteeing a multiplicity of solutions, if those solutions can indeed be found.

This is where a new axiom of choice is needed that will not automatically reduce the possible solutions to some unique set. On this score, the dialectical principle nicely fits the bill. It simply posits that the state of a process after some change is to some degree, or in some sense, "opposite" to – in the sense of different from – the state of the process before the change, and that change was impelled internally by the emergence of some contradiction. If the progress of the process is reconstructed and represented as the sequence of contours appearing at each and every change-of-state in the process, what has then been mapped may be considered as the stage-by-stage unfolding of each of the contradictions to which the progress of the process itself gives rise and their resolution. From this one assembles a sequence of testable hypotheses, and careful experiment can then establish what is likely by eliminating what is false or unlikely.

Since ancient times many phenomena of social as well as natural development have been presented in this light, as a "struggle" of opposites. The great knock against this approach always was that it seemed to displace any concern for the long-term, which in some respects exists in the present only in some idealized form, with immediate or short term concerns of the "here and now". In formulating the groundwork for our "economics of intangibles", however, we have found this approach reopens important questions that the evidence of events have proven were in fact not settled or explained satisfactorily by existing theory, and in a manner that does not allow the loop to be closed before all accumulated relevant elements of knowledge have been examined and applied to explaining and accounting for the development of whatever the phenomenon of interest. In and of itself, the dialectical method may be applied to explaining phenomena of the external material world to satisfy a hankering for the short-term solution or to reposition what is happening in the present in terms of what is best for the long term. Which direction is something that depends on the intentions of the investigator / researcher.

4.2.5. Placing t_{LINEAR} on Life Support

At the level of consciousness and thinking about these large social questions, persistent efforts to extend the lifetime and lifeline of t_{LINEAR}-based methods and lines of research in the social sciences continue to pose a major obstacle. The distorted view perpetrated by the contributions to the *U and SM* volume would not have been possible to engender otherwise, let alone continue to recapitulate itself a generation later into a new century (Petras and Veltmeyer 2004).

The cribwork for the theoretical foundations of the work of the contributors to *U and SM* came from three distinct but related sources, each of them a variant on the common theme of how to describe and realistically render, using modeling based on t_{LINEAR} conceptions, the *dynamics* of economic growth based on private accumulation, *i.e.*, how the conventional capitalist model moves through space and time. The first variant of this one, common t_{LINEAR} –based dynamics was elaborated in the work of the Stanford University economist Paul A. Baran entitled *The Political Economy of Growth* (Baran 1957); the second by a German-born scholar who was one of the earliest U.S.-educated "Sovietologists", Andre Gunder Frank, in his classic work *Capitalism and Underdevelopment in Latin America* (Frank 1967) which acknowledged an academic debt to Baran; and the third in the work of the African-born and French-trained economist Samir Amin beginning with his *Accumulation on a World Scale* (Amin 1970) which acknowledged a debt to the work of Baran, his long-time associate Paul Sweezy and Gunder Frank.

Baran himself marks the starting point of this work on the dynamics of capital accumulation explicitly in the Introduction to the Second Edition of his book released in 1962 by pointing to the influence of various developments in world politics on his decision to carry out this project. The key developments he mentions are the Bandung [Indonesia] Conference of 1955 in which the foundations of the present-day Non-Aligned Movement were laid (see *infra*), and the Twentieth Congress of the Communist Party of the Soviet Union in February 1956. In other words, this was a theory that acquired legs to the extent that it engaged some of the causes and consequences of changing directions in the socialist camp and stimulated theorisings about a "third way" outside capitalism or socialism.

On the latter occasion, the Khrushchev group ascended to full power throughout the Soviet party and state. When Khrushchev used the occasion to declare a great deal of the previous thirty years' political development null and void or a distortion due to excessive promotion by the former leadership of a so-called "cult of personality" around the person of Joseph Stalin, the former General Secretary of the party as well as Premier and head-of-state, he scandalised public opinion throughout the Soviet bloc, unleashing dangerous genies from various bottles which compelled Soviet military intervention against Hungary, a fellow socialist country and Warsaw Pact ally, in November 1956. Simultaneously and however indirectly, Khrushchev also gave encouragement to a variety of interventionist schemes of the U.S. and former European colonial powers, including most notably the Suez debacle of October 1956 initiated jointly by Britain, France and Israel (which ended with Israel becoming a client-state of the

United States), the collapse of the Fourth Republic in France and of the Eden government in London, the Anglo-American plots to assassinate the leaders of Syria and Iraq during 1957 and the subsequent US invasion of Lebanon in 1958.

Implicitly accepting Khrushchev's critique of Soviet economic and political development from 1924 to 1952 as definitive, Baran repudiates the entire politics and economics of socialist construction in the Soviet Union and asserts that the socialism of the Soviet experiment shares many of the deficiencies and marks of backwardness notable in many newly-independent or decolonising countries of Africa, Asia and Latin America but blames the massive and continuous external pressures applied by former European colonial empires and current imperial powers like the United States for causing such "underdevelopment" to persist in countries both socialist and capitalist (Baran: *ibid*).

The essential thesis put forward by those contributors to *U and SM* who were concerned to elaborate RCU theory is that entrapment in a state of permanent inequality as a consequence of "regional underdevelopment" gives rise to social movements seeking economic and political changes. According to their thesis, the role assigned to the construct described generally as "capitalism" is two-fold. First, in its modern form as both a generalised as well as global system of economic colonisation that is no longer confined, or available only, to this or that so-called Great Power, this "capitalism" comprises a set of relations that may entrap economic regions (which may include sectors of economies or the economy of entire countries or even groups of countries) in systems marked by more or less permanent social, economic and political inequalities. Second, it is in the nature of these inequalities that they are common to the capitalist social and economic system prevailing in the "metropolis", *i.e.*, in the region whence originated the investments in production in the underdeveloped region.

To what part, if any, however, of the real-world history of the notion of "underdevelopment" does this abstracted conception actually apply? As one looks around, in the present, at many regions of the globe, there are numerous struggles in which demands for economic and social equality, or an end to specific inequalities, are being raised. Does this tell us anything other than that we are living presently in a period that undergoes continual change, development and motion as societies and regions at different levels and stages of development sort out all manner of contradictions? It has the merit of demonstrating from actual facts on the ground the absurdity of the kind of steady-state equilibrium posited in more strictly conventional economic theories. As a description of present-day development, however, this conception of "regional underdevelopment", or even "regional capitalist underdevelopment" is not only unexceptional and uninformative, but it is also actually a starting-point for a great deal of disinformation.

The relevant modern real-world history of the notion of "underdevelopment" begins indeed with the conference referenced by Paul Baran. This was hosted by then-president Sukarno of Indonesia at Bandung, Indonesia in 1955 of 22 countries from Asia and 7 from Africa. The roster notably included one very large, and at that time anti-capitalist, regime, the People's Republic of China. Bandung's deliberations were also watched with interest from then-Soviet Russia, Central and Latin America and even France, a Western country with an extremely negative colonial past in Africa and Asia which had just suffered a serious military defeat in Vietnam, at Dienbienphu, the year before. It was very consciously snubbed and boycotted, on the other hand, by the United States and the other member states of the NATO alliance as being "pro-communist". The conference formulated the Bandung Principles, which would become the basis for establishing the Non-Aligned Movement in 1961 (the label of the grouping was supposed to define non-alignment of any member with either the United States or the Soviet Union, although in many cases this was more military-diplomatic fiction than economic reality).

The main thrust at the time, however, while not in fact "pro-communist", was clearly opposed to any continuation, either as "aid" or in any other form, of schemes for further plundering the natural resources of these countries on the basis introduced by European and North American colonising powers. These methods were blamed for preventing the economies of these countries from ever catching up with and fully providing for the full range of needs of their own populations. "Underdevelopment" was thus intended to describe both the current economic level of these countries and the future they faced if their current course did not change. There was no confusion whatsoever among the participants that a future hewing wood and drawing water for former colonial exploiters was no improvement whatever on their previous condition of direct colonial enslavement, and that economic development based on extracting and exporting raw materials without further processing and without using these raw materials to develop home industry offered a future without hope, *i.e.*, no future at all worthy of the name.

With the inauguration of the Kennedy Administration in 1961, the United States changed course in the policy area of foreign aid. This was widely justified and rationalised as a "liberal" swing of the pendulum back from the conservative extremes of the Eisenhower administration. However, as was clearly exposed by the American adventure at the Bay of Pigs, which militia-level people's forces of the Castro regime repelled and smashed with little need of heavier artillery support from the regular armed forces, the political essence of the Kennedy administration remained no less reactionary than its predecessor. Similarly, the aims of its foreign aid programs also remained the same. However, the effort became invested with a new justification in the theories of "economic takeoff" (Rostow, 1960).

The presentation of this theory married a discredited notion from the Victorian industrial era – that of the "deserving" and "undeserving" poor – to a strikingly modern idea, borrowed from atomic physics, of the so-called "critical mass". Proposing that U.S. foreign aid should be increased to many areas of the world up to then ignored – but mainly for purposes that would stimulate private investment and markets for U.S. goods ("deserving poor") as opposed to subsidising governments' ability to subsidise non-profit, not-yet-profitable or unprofitable necessary social services ("undeserving poor") – the Rostow model contended that such selectively targeted "aid" would assist those countries already enjoying a certain "critical mass" of private-sector-based economic development in the private sector to reach "economic take-off" and grow their way to a "modern" economy (and high-consumption "Western" way of life) for their citizens.

For the U.S. economist Andre Gunder Frank and his co-workers, it was increasingly apparent that this "aid" bolstered Latin American dictatorships in power against, and at the expense of, their own people. Even disseminating aid to "the deserving poor", so to speak, *i.e.*, on the basis of "takeoff" criteria being fulfilled, could only benefit a tiny elite at the top while continuing to condemn the vast majority to severe impoverishment and "underdevelopment". Hence, these analysts concluded, there was a deeper structural problem, or set of problems, which would have to be addressed in those societies and that could not be solved in principle by outside aid – no matter how free of strings. (Gunder Frank 2000)

Here, then, originated the theory of *regional* underdevelopment taken up by the contributors to *U and SM*, which will be tackled *infra*, as well as the debate among a number of variants, *e.g.*, "development of underdevelopment" and "uneven and combined development".

The concept of "underdevelopment" presented by Baran, Gunder Frank and Amin is derived from a peculiar theory about the nature of economic development. It is a theory based on recasting the rise of the Soviet economic model as an alternative variant of one and the same paradigm of Western economic development since the Renaissance. They achieve this identification by taking one feature of the Soviet economy and absolutising it. The feature they absolutise is the orderly intervention of the State in regulating the sphere of operation of the Law of Value, by, for example, subsidising the supply on the one hand of necessary goods or services while surtaxing revenues generated from the sale of luxury goods or services.

This absolutisation does two things. First and foremost, it dismisses or ignores any role for revolutionary struggle. The Bolshevik Revolution uniquely and alone provided the energy to eliminate the old Tsarist state, bureaucracy and army and expropriate the entire property of the foreign and big-Russian owners of heavy industry. Yet, obviously, without such an intention, no such category as "Soviet economy" was possible or conceivable. Such marginalising and narrowing of focus simultaneously set aside any discussion of the revolutionary aim of the Soviet system on the economic front – which was not "development" in the sense of accumulation of capital but transformation of the very relations of production and social life based on eliminating private ownership over the means of production.

Second, such absolutising the planning and interventionist features of the Soviet system created the impression that the only important difference between Soviet and non-Soviet economic systems was the lack of state planning in the non-Soviet systems (Baran: *ibid.*). That argument runs something like this: because their governments do not step in and regulate, wealth and poverty in non-Soviet economies accumulate at opposite social poles, thus piling injustice atop social inequality. So, although Soviet-type societies on the other hand cannot overcome inequalities due to natural differences in talent, etc., their state intervention attenuates any tendency towards injustice. The problem here is that the lack of state intervention in the economy of a non-Soviet system predicts absolutely nothing whatsoever about the level of societal justice or injustice. Hence, the conclusion that, if the state is looking after the people's economic needs, its intervention also becomes a force for increasing social justice is unwarranted.

This was the period of the politics of the Cold War. In that era, who was going to show public contempt towards such avowed well-wishers even when their proffered "help" was really unwelcome? In any event, no one at the time in the Soviet Union said such things about their own system. Its implication was demonstrably false and the reasoning that produced it deeply flawed. If what these writers had absolutised indeed constituted the *principal* difference between Soviet and non-Soviet economic systems, then the difference between the two systems would reduce merely to one of policy objectives. Since each member of every policy-making community wants only what is best for their own people, the exercise degenerates into a stale argument about matters that can never be decided. In actual fact, this so-called "socialist-capitalist convergence", as it was then called, was converted into so-called "peaceful competition." This was the context in which Khrushchev's group absurdly promised the Soviet people would catch up to and surpass the United States by 1970. (At the time, the following bitter joke circulated widely against this commitment to forget about any further revolutionary or qualitative transformation of Soviet life: "under capitalism Man exploits Man but under 'socialism' it's the other way round.") The aim, meanwhile, of these theoretical acrobatics in the model-building exercises of Baran, Gunder Frank and others following them was to create a typology of human progress. This typology would anchor comparisons of human social progress in general, including level of meaningful economic development, or relative underdevelopment, across different societies. It would rank societies according to how large or fast-growing the accumulation of poverty and other negative phenomena might be. From this typology and ranking, information could then be assembled into a realistic picture of the true relations between the exploited and the leading social classes of each society.

One consequence of this evolution in the field of economic theory, the promotion of a so-called "third way" between capitalism and socialism, has been widely discussed elsewhere (Blair 1998). Our interest here, however, is to deconstruct the "development-underdevelopment" continuum on which the entire subsequent evolution in international economic theory and practice came to be based. All kinds of societies could now be ranked and compared on one and the same "development-underdevelopment" continuum – Soviet-bloc countries, developing countries and developed countries. What Baran, Gunder Frank *et al.* called "underdevelopment" and defined as development's polar opposite was in fact, however, *not* just the *negation* of development as they were suggesting, *i.e.*, not just the accumulation of excessive poverty at one pole. Furthermore, despite the appearance of a potentially universal range of application to Soviet and non-Soviet, developed and underdeveloped, this apparent broadening of the field-of-vision was actually a narrowing achieved as a result of chopping the role of revolutionary, transforming struggle out of the picture. What they called "underdevelopment" was actually a subcategory of a much broader, but partially intangible, idea of interconnections and disconnections between the growth of tangible material forces of production and immaterial or intangible relations of production (Wallerstein 1974).

The very different effects of narrowing or widening a field of definition can be quite dramatic. Compare what is involved in mathematics, for example, when the derivation of formulas for "sin $n\theta$" or "cos $n\theta$" is attempted *without* any knowledge of the complex-number field, to what is involved *after* such knowledge is acquired. In the former case, restricted to the real-number field, quite elaborate plane-geometric or Cartesian-coordinate figures are required, considerable symbolic computation is involved for each different positive integer value of n, and the sin $n\theta$ and cos $n\theta$ formulae have to derived separately each time. In the latter case, simply by broadening the view-plane to the complex-number field, in which a number of the general form "cos θ +$j$sin θ" can be represented by the exponential $e^{j\theta}$, it becomes a matter of raising such a number to the n-th power, or $(e^{j\theta})^n$ = (cos θ +$j$sin θ)^n; rearranging as $e^{j(n\theta)}$; re-stating (cos θ +$j$sin θ)^n = cos nθ + $j$sin nθ as a binomial expansion of (cos θ +$j$sin θ)^n; and, finally, collecting all the real-valued terms as the equivalent of cos nθ and the imaginary-valued terms as the equivalent of sin nθ. Instead of struggling asystematically with geometric figures, an algebraic expression expanded, for any chosen value of n, according to a known formula in order to provide the necessary sequence of coefficients gives the result simultaneously for cos nθ and sin nθ.

Proceeding from its safely narrowed field-of-vision, shorn of dangerous "revolutionary baggage" and "rhetoric", the notion and theory of "development" in its original most pristine form – that laid out in Baran's *Political Economy of Growth* (1957) – comes asymptotically close to the underlying truth that would expose just how alien from nature and history the temporal notions embedded in conventional capitalist development model are... only to diverge at the last possible moment. Thus, Baran correctly distinguished the concept of surplus from the notion of profits. However, he completely missed the serious and essential, even defining, difference of temporal dimension involved: "profits" are associated with t_{LINEAR} whereas "surplus" is associated with $t_{HISTORICAL}$. What does this mean? What is its significance?

In order to grasp the economics of intangibles, it is necessary first to appreciate that fact that t_{LINEAR} is not the same as $t_{HISTORICAL}$. With the latter, cycles reappear, but nothing can exactly repeat because context was changed by development during the previous cycle or since: this is exactly what happens with the social surplus, which cannot and is never intended to be consumed in a single cycle or accumulated to some final value after some finite passage of time. With t_{LINEAR}, on the other hand, differences arising from mere temporal displacement of subsequent cycle(s) of similar development(s) are less consequential than structural similarities – sometimes even much less. Thus t_{LINEAR} is an essential instrument for measuring and-or predicting the profits generated by a particular but cyclically-repeated production arrangement – a structural similarity recurring in each cycle – involving some given quantum of capital advanced as wages and some given quantum of capital being exhausted in the wear and tear of equipment and consumption of raw materials by the production process of that cycle.

From the standpoint of t_{LINEAR}, it seems logical and possible to argue thus. A European cultural, economic and political setting framed the emergence and development of industrial capitalism and its mode of capital accumulation in a

European setting. A similar pattern and degree of development was not achieved when industrial capitalism came to parts of the world outside western and central Europe. It would seem to follow that, in attempting to account for one and the same "development / underdevelopment" nexus throughout the capitalist system anywhere on the face of the globe, the required categorisations and narrative are incomplete and require supplementation. What made sense of, and sustained, a coherent critique of industrial capitalism and its mode of capital accumulation in a European setting cannot provide a common meaningful narrative for all examples of development / underdevelopment throughout the capitalist system on world scale. For this purpose, t_{LINEAR} is neither the correct nor applicable temporal factor. The appropriate temporal factor to employ in all such cases is $t_{HISTORICAL}$: once development of industrial capital accumulation starts to take place, the context changes in both the hinterland and the metropolis, and the next locale in which development / underdevelopment appears will now operate somewhat differently.

(Of course: the data of contemporary and historical events and development could be more readily handled if everything were reducible to t_{LINEAR}. The problem is that such reduction, however precise and seemingly complete and closed within the terms of its own scale, also entails a loss of information at other scales. Could this information be preserved from loss by incorporating a $t_{HISTORICAL}$ metric? The following line of argument is certainly suggestive. Obviously: $_2t_{LINEAR}$ - $_1t_{LINEAR}$ = Δt, at some scalar value, whereas it is difficult to define what computing "$_2t_{HISTORICAL}$ - $_1t_{HISTORICAL}$" might mean, and at the same time $t_{HISTORICAL}$ is a far less trivial notion than t_{LINEAR}. One approach might be to define a quantifiable entity called $\tau_{HISTORICAL}$, comprising a "real" t_{LINEAR} component and an "imaginary" component labeled "β" which is a composite index incorporating some quantifier of how long the current historical cycle-of-interest has lasted, some qualifier of the historical sub-period, and quantifiers of the number of characteristic features of the sub-period that have persisted and that have disappeared respectively. Thus $\tau_{HISTORICAL} = t_{LINEAR} + j\beta$, which could be expressed (and more readily manipulated) as $\tau_{HISTORICAL} = \exp[j*\arcsin \beta/(\sqrt{t^2 + \beta^2})] = \exp[j*\arccos t/(\sqrt{t^2 + \beta^2})])$

Outside those countries actually wrestling with constructing and sustaining a socialist social economy with the fullest participation at all levels of society in the tasks confronting the entire society – Cuba today, for example – the line of march on which Marx set out has been ignored by academic economists. Marxian social science was deemed value-loaded, biased against private property and consumed with pursuing a single-minded political agenda (any of which is true for those who consider the *status quo* all-important). Most academic economists abandoned further efforts either to refute Marx's method or otherwise deal seriously with it.

Others, however, like the theorists of "underdevelopment", have wrestled with reconciling Marx's uncompromisingly $t_{HISTORICAL}$ approach with the t_{LINEAR} approach drummed into their consciousness and practice from formal training in academic social science. The contributions to *U and SM* by their acolytes typify the eclectic upshot of such attempts to reconcile the irreconcilable. This became yet another direction from which efforts would be launched to maintain t_{LINEAR} and its legacies on life-support.

From a $t_{HISTORICAL}$ standpoint, the increasing replacement within the conventional capitalist economic system of living labour by dead labour, *i.e.*, automation and the microcomputer, must outstrip the generation and origination of surplus from the exploitation of living labour. From this, Marx elaborated his theory of the falling rate of profit as the basic tendency of this economic system. However, examining the capitalist order headed by the United States in the late 1950s and early 1960s, following an extended period of economic growth in that country without major recessions or depression, Baran and his co-author Paul Sweezy (Baran and Sweezy 1966), who would outlive him to see their joint work, *Monopoly Capital*, into print, noted a continuing high level of profit alongside a rising surplus. From this he concluded that Marx's prediction of a basic tendency of the rate of profit to fall was a double-barrelled mistake. First, while this tendency was observable in the industrial system of mid-19th century Britain, it engendered an over-enthusiasm about revolutionary prospects left unfulfilled by subsequent events. Second, the industrial and financial structures of capitalism that emerged after Marx' and Engels' time eliminated the tendency, negating the entire line of theory developed around it.

The flaw in this line of reasoning, however, lies in its basis, which is the t_{LINEAR} approach. What actually happened was that the organisation of new wars, especially world wars between rival imperialist groupings and cartels, greatly supplemented the generation and origination of surplus from the exploitation of living labour sufficiently to override for entire periods the increasing replacement of living labour by dead labour. Hence the contradiction and its basic tendency persisted even if punctuated by periods during which the expanded sources of surplus through extra-economic means overwhelmed the normal operation of the economic law. However, Baran and Sweezy, fanatically and ideologically predisposed to push $t_{HISTORICAL}$ firmly and finally off a cliff and proclaim "the Way, the Truth and the Light" of the t_{LINEAR} approach, concluded instead that the rising surplus in and of itself had become the dominant tendency. As for pooh-poohing predictions of revolution, has it turned out anywhere that this economic system changed gears and started producing greater satisfaction and less want? On the contrary: as long as the exploited class, which is this system's special and essential product, did not rise up, this system would and did develop into one that "eats its young", a system of monopolies and cartels that fleece the peoples at many levels and on a world scale. Again: the t_{LINEAR} – minded, who hoist the telescope to look through the wrong end, cannot account for such an evolution.

One implicit thesis of the development/underdevelopment eclectics is that if there is not a non-linear branch-point event supplied by something like revolutionary overthrow, then there will no other further branch-point. From a $t_{\text{HISTORICAL}}$ standpoint, on the contrary: if there is not one non-linear event, *e.g.*, revolutionary overthow, there will be another non-linear event *viz.*, displacement of free competition by monopoly. Thus, *e.g.*, the predictions, by English Fabian socialist economists – starting before World War I with Wicksteed (1910), a fan of Jevons' work on marginal utility, and going all the way up to the Webbs (1920) and G.D.H. Cole (1944; 1956) after World War II – of smooth gradual transformation being averted by the peaceful reformist path is only possible by assuming a t_{LINEAR} path. Looking through the wrong end of the telescope, the devotee of the t_{LINEAR} view sees that there is a rising surplus alongside rising profits, but – lacking the depth of view available to those who assess these matters from a $t_{\text{HISTORICAL}}$ standpoint, does not grasp that these can only be consumed by some destroying the capital of others. Baran saw in the ever-rising surplus the signs of waste and parasitism, *e.g.*, the entire military-industrial complex, but the necessity of U.S. subordination of capitalist competitors in order to keep that surplus rising seems to have escaped his ken. Neither he nor Sweezy would ever connect the dots. This could have clarified that only the surplus in the dominating imperial centre rises without limit and only for so as long as it is "on top".

4.2.6. Merchant's Capital – Key Historic Intangible of the East Coast Fishery

It is undeniable that there has been great social and economic backwardness in the Canadian Atlantic provinces, largely as a legacy of British colonial rule and its articulation of an economy that engaged not merely in primary production but in producing outputs for end-markets tailored to requirements set by the British colonial system, controlled entirely from outside. Many backward-looking social relations and conditions were retained especially tenaciously in the fishing outports even for centuries. The problem is not with the accuracy of describing such phenomena as examples of "underdevelopment," but rather with using this concept to explain away everything. This tendency emerges directly from the reduction of all phenomena, especially in the work of Gunder Frank (1969), to some location on the all-embracing "development / underdevelopment" continuum. The concept becomes so broad as to end up explaining precisely.. nothing.

The Baran-Sweezy conception of monopoly capital which supplements and informs the theory of the "development / underdevelopment" continuum is riddled with many unstated, unwarranted and highly contradictory assumptions. Some are more fundamentally erroneous than others. For example, Baran and Sweezy certainly seem to subscribe to the notion that causes and effects in social-economic systems are objective processes taking place independent of any individual's will. However, they also posit that such a social-economic system of nominally material cause and effect may be driven by arbitrary and-or random intersections of the will of the Giant Corporation with the operation of these laws. In other words, some of the driving forces of this system lie outside its own laws. This is very much like the theological view which affirms the universality of Newton's laws of motion while also affirming that a Deity must exist outside time and space capable of intervening in and possibly altering these laws or their operation. Making this special allowance for corporate deity, the "critics" published in *U and SM* end up espousing the same metaphysics as the conventional economists they oppose. They differ with the conventional economists' assertion that the Sun must revolve about the Earth, *i.e.*, that the laws of economic science are the creature of the will of Giant Corporations, by trying to accommodate this alongside the alternative possibility that how corporations work is a function of economic laws, *i.e.*, that the Earth revolves about the Sun.

This concession is critical. It amounts to asserting there can be phenomena within a system supposedly governed by objective laws that cannot be accounted for by the normal operation of these laws. Either analysis of phenomena is carried out on the understanding that the phenomena under study are accountable in terms of objective laws outside anyone's will, or else the effort reduces to just another exercise in stating an opinion dressed up as analysis but corresponding to nothing objective. The philosophical position underlying the idea that a system is entirely explainable in terms of the objective operation of its laws of motion independent of anyone's will is materialism. The idea that, on the other hand, within this system, there can also be unknowable things-in-themselves is Kantian idealism. The idea that both concepts can fit together within one and the same system is pragmatism. Pragmatic approaches are very appealing for quick fixes often beloved by engineers and economists alike, but they are deadly for the kind of theoretical understanding required for serious science.

The Great Corporation of fishing enterprise, as a thoroughly tangible object, was the embodiment of a number of significant intangibles that provided the driving force. From the time the Europeans arrived at the end of the 15th century, the east coast fisheries were constructed and prosecuted directly and specifically in accordance with the requirements of colonial policy and imperial ambitions. Innis, the key source for the much criticised "staples theory," was hitting the nail on the head and proceeding from the correct starting-point for investigating the development of the east-coast fisheries when he observed that "the fishing industry of the North Atlantic has been exogenous in its development." (Innis 1954). Linked crucially to this was the overweening role exercised by merchant over the actual fishery producers, a condition that would introduce habits of subordination tending to render the producers unfitted to defend their own interests effectively when outside interests directly threatened their livelihood. Merchant's capital was the most powerful intangible factor guiding the fate of the fishing communities of the region well into the 20th century – long after its former highly tangible economic role had become thoroughly marginal.

Starting in the middle of the nineteenth century, railway expansion greatly expanded the base of and investment in agriculture, forestry and other resource extraction industries. The industrial capitalist system of the time seized dominant positions throughout these sectors of the economy, as well as to others linked to them in Canada before and after Confederation. This largely wiped out the retarding effects of merchant's capital. In the east-coast fisheries, however, merchant's capital persisted in its distorting and destructive role well into this century because of features peculiar to the historical development of the fishery. The coastal fisheries and offshore fisheries originated with the rise of the capitalist mode of production during European colonial expansion to the New World, India and China. The creation of a world market vastly expanded the basis for commodity exchange. The rise of the capitalist mode of production stimulated manufactures and further development of the division of labour as the feudal system was increasingly breached.

During this phase of capitalism's ascendancy, the leading role was played by merchant's capital, the specialist in exchange. Capital in this form opened up the complex traffic and exchange of various raw and finished products as well as slave labour between Europe and its colonies and among the colonies proper — for example providing food for slaves from salt fish in Newfoundland, the rum-running of merchants from colonial New England between the West Indies and the Thirteen Colonies, etc. At Chapter XX of Volume III of *Capital*, Karl Marx explains that:

"Merchant's capital, when it holds a position of dominance, stands everywhere for a system of robbery, so that its development among the trading nations of old and modern times is always directly connected with plundering, piracy, kidnapping slaves and colonial conquest." (Marx 1892)

The capitalist mode of production arrived in the region of the New World colonised by England with the development of seasonal fishing enterprises by European fleets off Newfoundland at the end of the 15th century. However, as Marx explains:

"Merchant's capital does no more than carry on the process of circulation. Originally commerce was the precondition for the transformation of the crafts, the rural domestic industries and feudal agriculture into capitalist enterprises. It develops the product into a commodity, partly by creating a market for it and partly by producing new commodity equivalents and providing production with new raw and auxiliary materials, thereby opening new branches of production based from the first upon commerce, both as concerns production for the home and world market." (Marx: *ibid.*)

From the start of the 16th to the middle of the 18th century, merchant's capital played an important role in opening up the New World by virtue of its dominating the rise and development of the east-coast fishery:

"The merchant establishes direct sway over production. However much this serves as a stepping-stone.., it cannot by itself contribute to the overthrow of the old mode of production, but tends rather to preserve and retain it as its precondition." (Marx: *ibid.*)

In other words, this activity at its outset was propping up the Old World as much as it was opening up the New. Aspects of relations of production from the decline of the feudal system were transferred into the east-coast fishery from the outset. The methods for drying and salting fish catches, such as the so-called green cure which came out of the feudal system in Brittany. The methods for paying and hiring fishermen on the basis of so-called "catch shares" and "boat shares" similarly came out of late-feudal Europe. These methods were not yet fully capitalist. Unlike the proletarian, the fisherman was not without some means of production. The merchant, unlike the factory owner, "shared" the means of production because he could not fully own them. However, just as it was the merchant's dictate that set these relations in motion, it was also the merchant who was in the position to bind fishermen to him by advancing credit against future production. Marx points out that:

"This system everywhere presents an obstacle to the real capitalist mode of production and goes under with its development. Without revolutionising the mode of production, it only worsens the condition of the direct producers, turns them into mere wage workers and proletarians under conditions worse than those under the immediate control of capital." (Marx: *ibid.*)

On the world scale, "as soon as manufacture gains sufficient strength and especially large-scale industry, it creates in its turn a market for itself, by capturing it through its commodities. At this point commerce becomes the servant of industrial production, for which continued expansion of the market becomes a vital necessity." (Marx: *ibid.*)

The mercantile system and merchants dominated the east-coast fishery from the end of the 15th century to the middle of the 18th, before commerce would "become the servant of industrial production." In this period, the settlement of the fishing areas along the eastern seaboard was severely restricted, and outrightly forbidden in Newfoundland, by merchant's capital. As large-scale manufacture arose in England, converting merchant's capital into its servant, the population of North America proceeded to expand. This provided the market that would be captured by English manufactures in America during the earlier part of the nineteenth century.

In the fishery of its remaining North American colonies, however, after the Anglo-American colonists won their independence and with the rise of industrial capitalism in England, the yoke of merchant's capital over the east-coast fisheries was intensified. This form of capital was uniquely positioned to link the colonial territories as a market for finished commodities from industry in England. This was encouraged insofar as it helped to keep British North America out of the clutches of competing interests from the United States. Commerce conducted by merchant's capital under these conditions "will have more or less of a counter-effect on the communities between which it is carried on. It will subordinate production more and more to exchange value by making luxuries and subsistence more dependent on sale than on the immediate use of the products. Thereby it dissolves the old relationships. It multiplies money circulation. It encompasses no longer merely the surplus of production, but bites deeper and deeper into the latter, and makes entire branches of production dependent upon it." (Marx: *ibid.*)

However, precisely what this "disintegrating effect" would be, and the forms it would take, depended "very much upon the nature of the producing community," according to Marx. That was why in the Thirteen Colonies, where capital was accumulated independent of the British, certain typical features of social disintegration did not appear which would on the other hand become rife throughout British North America in the early decades of the 19th century. In the fisheries of Newfoundland and the Maritimes, whenever an industrial or commercial crisis broke out, there were outbreaks of famine, disease, riotous and spontaneous uprisings of producers against the material conditions and - above all - massive emigration to the New England states in search of work.

The ruinous consequences of merchant's capital retaining its yoke long after it had exhausted any remotely progressive social role can be illustrated by comparing what happened to the popular impulse towards independence in the United States, in the British North American colonies outside Newfoundland, and in Newfoundland. The impulse towards domestic manufacturing grew like an incubus within the Thirteen Colonies, fuelling the anti-colonial independence war that would eventually give birth to the United States and making the social class interested in furthering this development its principal social beneficiary. In what would eventually become Canada, on the other hand, that same social class furthered its interests by signing away any and all rights or notions of genuine independence and national sovereignty in exchange for a protected position as a British Dominion, *i.e.*, as the world's first modern neocolony. In Newfoundland, however, the yoke of the merchant classes remained unchallenged by any local manufacturing interest. They retained their position and wealth by liquidating and diverting the slightest tendency among the people towards independence.

What the merchant did to the small fisherman in Newfoundland for about 300 years was done to small fishermen in the Canadian Maritimes on a less brutalising basis but with certain important similarities. In all four Atlantic Canadian provinces, the dependence of coastal communities on commercial fishing grew as a function of the entrepreneurial classes' freedom to compel relatively excessive numbers of people to remain involved and connected to the fishery as a source of income, especially part-time, and tied to the middleman either as the holder of the mortgage on some fishermen's boats, or as a supplier of gear or as the factor for getting the catch somewhere somehow into the market. By maintaining the coastal fisheries as a pool of cheap surplus-labour offering itself under terms of voluntary servitude, without overt external compulsion, the commercial operators in these fisheries tied up almost no capital of their own in equipment or wages for any extended length of time.

The salt fish trade was the mainstay of Canada's and Newfoundland's Atlantic fisheries. Whem it collapsed after the First World War, new products had to be produced by modern fish-processing factories for markets in the United States. The mercantile interests of the previous period and its arrangements weathered the transition largely intact by interposing themselves between the working fishermen from the Canadian coastal communities and the foreign investors, mostly from large U.S. food-processing corporations. (The latter, starting in the 1920s, were establishing processing plants along Nova Scotia's South Shore.) Thus, for example, local fisheries middlemen claiming to speak in the name of the coastal fishermen opposed the entry into the Canadian east coast fishery of large scale fishing trawlers from New England. Naturally the Canadian coastal fishermen were indeed concerned but few if any could know the extent to which a number of these self-appointed saviours were already either themselves fronting for, or working with the locally-organised corporate combine fronting for, the U.S. fish processing interests in the province some of whom were already heavily invested in the New England trawling fleets coming to Nova Scotia waters.

Up until the groundfish moratorium of the early 1990s, all the Canadian-owned but vertically-integrated fishing companies, *i.e.*, owning their own trawler fleets supplying their fish plants, entirely based and fishing along the Atlantic coastline, as well as the smaller processors relying on independently-operating fishermen to supply them, were keen to maintain a large pool of surplus-labour in the fishing communities. By this time, unlike industry in the interior of Canada, the fishery received absolutely no new entrants from foreign worker immigration to Canada while its immediate workforce continued to age and the young generation in these communities, increasingly able to access the large cities, drifted away from following their father's or grandfather's career in the most unsafe labouring occupation in the country after coal mining.

To the contributors of *U and SM*, armed with their RCU theory, these cheap surplus-labour pools in the outports symbolised "underdevelopment." In later writings, some of them explicitly linked its persistence to the excessive fishing effort for which subsequent fish stock depletion was blamed and the eventual groundfish moratorium even justified. We now know, however, that while overfishing may have provoked officials to take a moratorium option seriously, the readiness of the corporate sector to acquiesce in this measure had nothing at all to do with enabling regional fish stocks to rebuild. On the contrary, when the Soviet Union collapsed in 1991, its fishing fleet – then the largest in the world – was largely liquidated by the Yeltsin government. Enormous surplus inventories of fish catches from the Russian fleet piled up on landing wharves literally around the globe ready to sell at distress prices. The Canadian moratorium enabled the largest fish companies, saddled with expensive fleets catching diminishing quantities of raw material per unit effort, to dump their groundfish fleets and supply all their own customers and markets by purchasing very considerable lots of these Russian inventories. The largest vertically-integrated Canadian fishing companies, who had depended heavily on their own fleets' catches, reduced their scale of operations by shedding much of their fleet operations but did not not lose money in net terms after the moratorium was introduced.

As part of the rules under which the Uruguay Round of the General Agreement on Tariffs and Trade restructured itself into the World Trade Organisation in the 1990s, the Canadian government was compelled to end many programs of direct subsidies to industry, including the construction of new fishing trawler fleets. Thus once the moratorium was in place, the leading vertically integrated processors had a strategy in place whereby they would return to their pre-1960s form as merchandisers of others' catches, only now the catches could be from anywhere around the globe and the foreign fishing fleet might land their catches in a Canadian port for processing. In fact this sparked a spontaneous rebellion by southwest Nova Scotia fishermen in July 1993. For more than a week, they blockaded a Russian fisheries vessel as it attempted to unload catches at a plant in Shelburne, NS. As there would be no further subsidy for constructing a new fleet, these companies also needed the moratorium to remain in place indefinitely as an argument against acquiring new vessels or rehiring fishermen. Thus the companies became invested in maintaining the moratorium not out of concern that the stocks off the east coast ever rebuild, but so that that their monopoly as globalised fish merchants would develop undisturbed.

The contributors to *U and SM* seeking evidences mainly and only of "underdevelopment", by and large completely bypass or miss any of these significances of merchant's capital in the east coast fishery. Given their approach as already described, and the blinders on their vision, where would they find it? In the 18th century Newfoundland outport, the merchant resided in St. John's or more likely in England. His agent might put in an appearance when the ship came at the end of the season to collect the dried product and settle up the accounts for the year. No cash whatever changed hands. Supplies of food, clothing and other things not producible in a settlement on barren rock devoid of resources plunked down in the middle a sparsely settled wilderness were issued "on account". The merchant and his "factor" controlled the accounting. In the Newfoundland fishery, the English cod merchant aristocracy managed to impose a state of degradation that, absent only the acts of open racist oppression, outdid the plantation slavery system of the American colonies in one significant respect: whereas the slaveowner still had the inconvenience of having to supply out of his own revenue his slaves' dietary needs, the English cod merchant told his indentured servants in the outports once a year that this was the state of the account, you owe me this much for food and supplies out of your labour which I have tallied, like it or leave town never to return. How much more tangible can the economics of an intangible become? (Morgan 1992)

The history of east coast fisheries both in Newfoundland and the Canadian Maritimes thus richly demonstrate how tangible and intangible *roles* cannot be confused with an economic category's tangible or intangible *appearance*. Exactly the same is true of commodity economy in general, as the next section discusses at some length.

4.2.7. The 800-Pound Gorilla

Is the enterprise or the commodity the basic unit of economic life? It sounds like an invitation to discuss how many angels could dance on the head of a pin, but in fact on the outcome hinges a great deal of misunderstanding and even disinformation about modern economic life. A correct understanding of all the phenomena of modern economic life, from how planes fly to how oil and gas are gotten out of the ground to refineries and residences and everything in between, depends on how this question is answered.

A principled analysis must uphold the commodity as the basic cell life-form of the capitalist economic system. That is where all the key intangibles reside. In that sense, the commodity plays the role of the proverbial 800-pound gorilla in the room: everyone knows it is there, none dare acknowledge it. This goes completely against the lines of analysis of both the conventional economists and that of their so-called "critics" contributing to *U and SM*. They each take as their starting point the operations and transactions of enterprise (firm or individual).

Chief among these "critics" is the Baran-Sweezy school and those such as the theorists of regional underdevelopment who derive their analyses from its positions. They see the capitalist system as an "immense accumulation of commodities" instead of penetrating the veil of the commodity-form. As a consequence they end up capitulating to commodity-fetishism, the religious reflex of the capitalist system, and baking their "theory of regional underdevelopment" as yet another version of the "theory of productive forces", *i.e.*, the idea that what people can do economically and politically is already pre-conditioned by, *and locked into*, the level of technologies and production already achieved.

The issue of the commodity as the basic cell-form of the prevailing economic system is a matter of considerable theoretical importance. Why take the commodity as starting-point of investigation? Firstly, commodity economy existed before there was industrial capitalism. It arose as the physical form in which exchange of goods can take place. Exchange arose to overcome the gaps and defects of division of labour in assuring adequate production and reproduction of socially necessary goods. Secondly, commodity economy can operate regardless of whether production is private or social. Our capitalist systems, however, are historically and structurally a special stage of development in the history of commodity economy, in two key respects:

   a) the issue and special circumstance surrounding commodities and capitalism is that only under the capitalist mode of production is labour-power bought and sold as a commodity and able to generate surplus-value only if bought and sold as a commodity; and

b) labour-power can only be bought and sold as a commodity if ownership is private and production is social.

Marx wrote in the first two sentences of the first chapter of the first volume of Capital:

> "The wealth of those societies in which the capitalist mode of production prevails presents itself as an immense accumulation of commodities, its unit being a single commodity. Our investigation must therefore begin with the analysis of a single commodity." (Marx 1867)

This is an extremely interesting choice for a starting point. It is right under everyone's nose. No college degree is required to grasp a commodity, and you certainly don't have to be a rocket scientist to notice commodities everywhere, forming that "immense accumulation." But what is most interesting about this starting point is that the bland and unremarkable surface appearance of commodities, taken individually or as an "immense accumulation," veils an extremely complex history and development of relations between exploiters and exploited.

In fact, the most deceptive feature of commodities, regardless how physically different or variable they are, is their very "thing"-ness. No matter how variegated the physical form, every commodity veils one and the same basic social relation — a civil war between Labour and Capital.

The commodity is the materialisation of value in society. Put another way – it is the tangible vessel for an entire array of intangible relationships between producers and Nature, the mother of all wealth, as well as among the producers themselves, whose collective labour constitutes the fatherhood of all wealth. This value can only be captured, extracted through exchanging commodities against each other. Furthermore - and this is peculiar to capitalist societies - every commodity contains unpaid surplus-labour which is also exchanged as commodities are exchanged. So: commodity-exchange means that, on average, value is exchanged for equivalent value, this value can only be realised through exchange - but, at some point en route, as the commodity was coming into existence, some surplus-value was appropriated by someone somewhere in this process.

The physical appearance of the commodity as a tangible object masks not only an entire array of intangible relationships: it also masks intention. The surface appearance is that there is no systematic or socially-organised compulsion for anyone to buy or sell commodities. In fact, one section - the workers engaged in actual material production - have, through prior appropriation by others of their means of production, nothing left but their labour-power to sell in exchange for the commodities necessary for sustaining their lives. They must exchange their labour-power as a commodity with the other section that owns the means of production. The surface appearance is that this exchange takes place freely, but in fact the means of production are privately owned by the same social interest that is uniquely positioned to purchase the workers' labour-power and dispose of it as a commodity. So the seller of the commodity of labour-power has first to submit to this law of private property. This dictates all the terms and conditions of the purchase and sale of labour-power as a commodity and commands all the fruits of the workers' labour. Unlike every other commodity, labour-power has no value in itself apart from its cost of reproduction. Yet it is the source of all other commodity values which accumulate in the hands of Capital as wealth.

Commodities have value in use as well as value in exchange. These use-values can only be realised when the commodity is consumed (purchased). However, such consumption presupposes production aimed not at producing use-values for their own sake but, on the contrary, solely at realising their value in exchange. Hidden in this exchange-value is surplus-labour appropriated by Capital at the point of production and realised (turned into money) as surplus-value at the point of exchange (through sale). So: under the capitalist mode of production and its labour-process, there has to be production in order to have consumption – otherwise, labour-power cannot be reproduced and sustained for resale. But, likewise, there must also be further consumption in order to have further production - otherwise, goods pile up unsold, and Capital ceases to realise surplus-value and ceases to accumulate.

Virtually by definition, however, the capitalist system separates, and indeed has to separate, the exchange-value of a commodity from its use-value. Otherwise, surplus-value could never be realised, and Capital could then not expand. This produces the conundrum whereby not only are production and consumption separated, but they are unable mutually to regulate each other, and instead of production developing smoothly, there are crises of overproduction from which Capital can recover only by liquidating an entire mass of productive forces (through layoffs, shutdowns, unemployment, the bankrupting and-or takeover of weaker competitors, rationalisation, downsizing, etc.). Crises are built in, not accidental.

The importance of identifying surplus-value is that it connects the added value to its source in living labour. The more common term for this surplus is "profit", but it is important to recognise that profit – meaning industrial profit, profits garnered from the organization of industrial commodity production – is but one branch of the entire social surplus. Rent and interest are also forms of surplus-value but they pre-date the rise of industrial commodity production and the generating of profits from the exploitation of waged, *i.e.*, living, labour. As a result, in general, economists and others fail to connect these forms of the social surplus with living labour. (In this sense, by the way, although profits are quite tangible, surplus seems somewhat more intangible.) The fact of the matter is that once industrial production became the main source of generating surplus in society, the rates of interest and of rent were adjusted to compete with industry in attracting capital for investment. Thus the rate of exploitation of living labour in industry becomes the trendsetter for the overall average rate at which rent is charged or money is loaned, and these rates move up or down with the industrial profit rate; the latter is a leading indicator of where the former will likely end up.

Thus, the operations and transactions of enterprise (firm or individual) cannot be taken as the starting point by anyone investigating how any section of an economy based on industrial commodity production and lead them to correct conclusions. The preceding is sufficient to establish that what happens in a sector such as the fishery in Atlantic Canada cannot be meaningfully understood mainly or purely as a function of arbitrary actions by either Big Government or Big Business (the Giant Corporation). A systematic process is at work, in which particular features of this or that company or government policy may provide specific content but cannot alter the basic form of the relations involved, or the essential result. Only by penetrating the commodity-veil can serious understanding of the theoretical issued be attained.

As part of yet another deeply intangible set of relations, the commodity-form veils how and why the social labour of individuals for others takes place under capitalism by representing "labour" as value, on the one hand, and labour-time by the magnitude of that value on the other. In the production and exchange of commodities, the real social relations of life and labour become disguised in the fantastic form of social relations between things, material objects, commodities. This is the signal that production under such a system has taken command over man. People no longer command their own productive activity or any aspect of how their labour-time is used. Indeed this is what renders "time" such a crucial intangible underlying all the categories of social-scientific investigation in general and economic science in particular.

On the one hand, the capitalist mode of production sets in motion a mechanism that compulsively socialises labour: people are compelled to produce for society and modern society lives at the expense of the actual producers, as opposed to the era of outright slavery when toilers lived at the expense of society.

Personal labour for others, which was clearly delineated in feudal society, disappears with the rise of capitalist commodity production. It renders all qualitative differences between different kinds of labour superfluous. It reduces all labour to an undifferentiated mass of social labour-time, and reduces the differences between kinds of labour to quantitative relations between this or that amount of labour-time – congealed in the form of the commodity and disguised in the value-form of the commodity.

The value of commodities, especially the surplus-value congealed as surplus-labour in the commodity is realisable only through exchange. Consciousness of how the capitalist system actually operates in this regard, however, cannot be gleaned without penetrating the commodity-veil. Left on its own, the real social relations of commodity economy and their potential remain wrapped in mystery. In such a society, commodities as such, in spontaneous consciousness, become society's central holy fetish or "religious reflex." This fetishising of commodities is peculiar to capitalism. To the extent that the producers themselves are not freely associated or conscious of the possibilities of becoming freely associated, they remain subject to apparently mysterious, apparently "unknowable," unconquered forces of Nature and society.

Consciousness is thus a function not of the level of development of the productive forces, or of whether a region or its people are "underdeveloped," but rather of the struggle the producers wage to free themselves from the social and economic fetters imposed by the interests of others on the lives they want to lead. The Baran-Sweezy school was only the latest in a long line to pay lip service to this definition of social consciousness and its source while proceeding along blithely to define the enterprise, such as the Giant Corporation, as the basic cell of the capitalist economy, rather than the commodity.

This spawns many irresolvable contradictions. For example, as previously mentioned, there is the case of the so-called "staples theory" of how Canada came into existence. This theory has been used to argue that the east-west character of Canadian national geography, the tradition of state involvement in the economic life of the country and thus the "essential" characteristics of the Canadians as a people arise out of how the fur trade enterprise or the cod fishing enterprise or the timber cutting enterprise or the wheat-growing enterprise opened up the northern half of North America.

The "Giant Corporation" approach leads to or is connected with many other dilemmas that paralyse real movement.

First of all, the target of any serious popular movement has been financial oligarchy. But the whole thrust of positing the Giant Corporation as the basic unit and engine of the system is to deny the existence and role of the financial oligarchy.

By the end of the nineteenth century, competitive capitalism had given way to monopoly in all fields. This monopoly capitalism became the basis of modern imperialism. Imperialism means not simply land-grabbing, or colonial policy or other processes identified only with particular countries and particular historical periods. Imperialism actually means a system and purpose for the entire social-economic order structured on a very definite basis, possessing a global reach and operating in all countries. It multiplies all the contradictions in various directions and adding to the basic contradiction between the bourgeoisie and proletariat other major contradictions such as that between nations oppressed by imperialism and the imperialists, and sharpening contradictions among the monopoly groups and between rival imperial interests.

Its economic base is monopoly, its political content is reaction all down the line, and it proceeds and spreads by way of local and world wars with war preparations as its most profitable business. However, Baran and Sweezy posit "something quite different." They locate the base of "monopoly capitalism" in "the giant corporation."

Initially they argue this does not negate the notion of the financial oligarchy, only the notion of its power and authority:

> "There is no implication .. that great wealth, or family connections, or large personal or family stockholdings are unimportant in the recruiting and promoting of management personnel. .. It may indeed be taken for granted that they are normally decisive. What we are implying is something quite different: that stock ownership, wealth, connections, etc. do not as a rule enable a man to control or exercise great influence on a giant corporation from the outside. They are rather tickets of admission to the inside, where real corporate power is wielded. Mills put the essential point in a nut shell:
>
>> Not great fortunes, but great corporations are the important units of wealth, to which individuals of property are variously attached
>
> "What needs to be emphasised is that the location of power inside rather than outside the typical giant corporation renders obsolete the concept of the 'interest group' as a fundamental unit in the structure of capitalist society. ..
>
> "A whole series of developments have loosened or broken the ties that formerly bound the great interest groups together." (Baran and Sweezy 1966)

In other words, financial oligarchy ('interest groups') where necessary, but not necessarily financial oligarchs: the 'giant corporation' is posited explicitly in opposition to the concept of finance capital as the deepest economic basis of monopoly.

Monopoly capital is an economic form. The giant corporation is another economic form. According to the logic of Baran and Sweezy, the latter is the innermost basis for the former. But, in reality, can one economic form be the basis of another economic form? The basis of an economic form cannot be some other economic form. What happens is straightforward enough: human animals, socialised independent of anyone's will, enter into definite relations for the purpose of reproducing existence — which is also independent of anyone's will. The sum-total of such relations form a social mode of production that characterises an entire epoch of social development. These social relations become crystallised in the form of definite social classes defined by their status within the given mode of production, especially as regards ownership or control over the forces of production. The activity of the members of these social classes according to definite relations of production give rise to an economic form. Nothing else can give rise to an economic form.

The giant corporation is a form of monopoly capital. As such it is an instrument of the system and rule of the financial oligarchy. The fact that a corporation may follow a course opposed to the desire of this or that financial oligarch or group does not mean that it is independent of the financial oligarchy. It means only that there are different competing interests within this oligarchy and one may have bested another. If there were no financial oligarchy there would be no giant corporations.

What is the significance of Baran and Sweezy's substitution of the giant corporation for the financial oligarchy? It is utterly pragmatic, *viz.*, "the location of power inside rather than outside the typical giant corporation." But what basis is there for asserting that the locating of power centres inside a corporation negates the possibility of power centres outside it?

The claim that the old industrial trusts ("interest groups") have been broken up is dishonest sleight-of-hand. For example, in the 1980s the American Telephone and Telegraph (ATT) trust, controlling the tens of thousands of electrical, electronic and telephone patents of the Bell group of companies, was broken up by court order into 13 "regional operating Bell companies" (RBOCs).[3]

---

[3] These companies, operated separately, nevertheless sent a single spokesman, from one of the companies that is now almost as large by itself as the entire AT and T system was at the time of the court-ordered breakup, to harangue the American president-elect Clinton in front of live gavel-to-gavel television coverage at his two-day economic summit in Little Rock, Arkansas on 15-16

By 2005-6, the wheel was coming full circle, with major RBOC spinoffs seeking or being sought for mergers with long-distance service providers, on the one hand, and unprecedented activities to merge entire telecommunications monopolies (combining control of wireless, land-line and Internet services) on transcontinental scale, in order to remain competitive in global market terms with individual telecommunication monopolies from India and China that service, within only one or two jurisdictions, hundreds of millions and even billions of customers. (Kermisch and Smith 2005; Silver, Young and Abboud 2006)

Baran and Sweezy's "whole series of developments" that "have loosened or broken the ties that formerly bound the great interests groups together" are evidence only for the limited proposition that alignments in the financial oligarchy can shift, not of some qualitative change whereby the power and role of the financial oligarchy has been terminated or displaced. According to the logic of Baran and Sweezy, the monopoly capitalist system suffers from incidental difficulties (some problem here or there), but there is no problem with this pragmatic method. The grand attack on the "power centres" of the "giant corporations" reduces to little more than an appeal to the corporations to reform themselves.

Is government intervention and involvement in the fishery incidental or fundamental? If it is fundamental, then the state machine must be understood as having been integrated under the sway of the monopolies. That is called "state monopoly capital." Another major defect in the concept of monopoly capital put forward by Baran and Sweezy, however, stems precisely from their repudiating the emergence, role and significance of state monopoly capital.

They write:

> "Lenin spoke of the 'epoch of the development of monopoly capitalism into state monopoly capitalism' . .
>
> "We have chosen not to follow this precedent but rather to use the terms 'monopoly capital' and 'monopoly capitalism' without qualification for two reasons. In the first place, the state has always played a crucial role in the development of capitalism, and while this role has certainly increased quantitatively, we find the evidence of a qualitative change in recent decades

---

December 1992. This industrialist's message was that the so-called "Baby Bells," as they are known, should be allowed to cooperate more openly precisely in the areas of computer telecommunications that provided the rationale for applying anti-trust restraints on AT and T in the first place a decade earlier. This "proposal" went on to be implemented as actual policy under a series of special committees directed by the then U.S. vice-president, Al Gore, advised by the former chairman of Apple Computers, John Sculley, one of the leading unofficial economic policy advisors to the Clinton administration.

unconvincing. Under the circumstances, to lay special emphasis on the role of the state in the present stage of monopoly capitalism may only mislead people into assuming that it was of negligible importance in the earlier history of capitalism. Even more important is the fact that terms like 'state capitalism' and 'state monopoly capitalism' almost inevitably carry the connotation that the state is somehow an independent social force, coordinate with private business, and that the functioning of the system is determined not only by the cooperation of these two forces but also by their antagonisms and conflicts.. In reality, what appear to be conflicts between business and government are reflections of conflict within the ruling class." (Baran and Sweezy: *ibid.*)

State monopoly capitalism means precisely that the state power has been subordinated to the interests of monopoly. State power is not independent of the power of the monopolies. Nor is it co-ordinate with, let alone competitive against, such private power. Furthermore, although the state has always played a role, even a "crucial role," in the development of capitalism, what is different under state monopoly capitalism is that the involvement of the state machinery becomes a norm. No longer is it just a particular intervention at a particular time.

What is qualitatively new about this state compared to the state under competitive capitalism is the narrowing of its social base of support and the narrowing of its perspective. After substituting "the giant corporation" for the financial oligarchy so as to make the crucial political role of the financial oligarchy disappear, however, it is hardly surprising that Baran and Sweezy should "find the evidence of a qualitative change..unconvincing."

Baran and Sweezy's repudiation of the notion of "state monopoly capital" is also based on an unsupported claim that the rise of monopoly may have altered the "laws of motion" of capitalism.

According to Baran and Sweezy, "the Marxian analysis of capitalism still rests .. on the assumption of a competitive economy" because, although Lenin pointed out that "imperialism is the monopoly stage of capitalism," apparently "neither Lenin nor any of his followers attempted to explore the consequences of the predominance of monopoly for the working principles and 'laws of motion' of the underlying capitalist economy" in which "Marx's Capital continued to reign supreme." (Baran and Sweezy: *ibid.*)

The evolution from competition to monopoly, however, takes place within one and the same capitalist mode of production. The laws of motion and working principles do not change for a given mode of production. Rather, if different working principles and laws of motion apply, the mode of production must have changed. The shift from competition to monopoly represents a shift not in the working principles or laws of motion, but in the motive for production under capitalism. It is a shift away from being satisfied with average profit to requiring and insisting on nothing less than maximum profit.

Baran and Sweezy misrepresent the effect of this change on the motive for production (from average profit to the maximum profit) as a change in the central tendency from the tendency of the rate of profit to fall into some alleged tendency for the surplus to rise forever.

This is the most outstanding and notorious distortion of their work. The cogent argument of Karl Marx was that the class aim of the capitalist system – to extract the socially average rate of profit in a given branch of material production on the basis of maximising the rate of extraction of surplus-value, which means simply to maximise the exploitation of the workers of a given enterprise or enterprises of a given capitalist – was at loggerheads with its central tendency for the rate of profit to fall. Contradictorily and inexorably, even while tending to push the system forward, this could trigger devastating crises. But Baran and Sweezy eschew the notion that the monopoly capitalist system has a class aim. By thus disregarding or disclaiming any class aim for their system, the contradiction between class aim and the central tendency has been suppressed.

Baran and Sweezy's central tendency amounts to declaring that monopoly capital has an infinite capacity to regenerate and eternally reproduce itself. Evidence of counter-tendencies such as revolutions and overthrow are considered aberrations. According to this logic, the problem is in the sphere of consumption, as opposed to the sphere of production, which begs the question of how the social relation of Capital (and monopoly capital) was created in the first place. Everyone knows the chicken cannot be separated from the egg, and as a practical matter the egg cannot be placed before the chicken. Similarly, there could not be consumption separate from production, and, as a practical matter, something — the commodity — must first be produced before we can speak about consumption. This illustrates how positing the analysis of the entire social-economic order of monopoly capitalism on the asserted central tendency of the surplus to rise without limit leads to various positions that are rife with other contradictions and errors. This leads to the idea, for example, that crises arise from and reflect not overproduction, but only underconsumption. Hence what needs attention is only the system of circulation and distribution (to ensure the "immense accumulation of commodities" is distributed and consumed).

By thus reducing the world to bite-sized digestible bits and eliminating the big picture, intangibles are reduced to something ghostly and evanescent, if not the stuff of conspiracy buffs, and the "reconciliation" thus obtained between t_{LINEAR} and $t_{HISTORICAL}$ appears little different in principle from the "reconciliation" achieved in the fairy tale between the lion and the lamb.

REFERENCES AND BIBLIOGRAPHY

Abbott, Edwin Abbott. 1884. *Flatland – A Romance in Many Dimensions* London: Macmillan.
Allen, David W., Richard B. Allen, Robert E. Black *et al.*1979. *Effects on Commercial Fishing of Petroleum Development off the Northeastern United States.* Woods Hole MA: Woods Hole Oceanographic Institute.
Amin, Samir.1972. *Accumulation on a World Scale.* New York:Monthly Review Press.
Amoroso, Bruno 2003. *Global Apartheid* Roskilde [Denmark]: Roskilde UP.
Anders, George. 2005. "As Oil Prices Swing, Gas-Station Owners Try Futures Market", *The Wall Street Journal* [New York] Jun 21 p.A1.
Anderson, Eric L. and Malcolm L. Spaulding.1981. "Application of an Oil Spill Fates Model to Environmental Management on Georges Bank" *The Environmental Professional,* 3:131-32.
Armstrong, Jane. 2006a. "Revival On The Rock- Williams Seeks A Piece Of The Action [Part 1 Of 3]" *The Globe And Mail Report On Business* [Toronto] Mar 15 p.1.
———. 2006b. "Black Stuff Puts Nfld In The Black", *The Globe And Mail Report On Business* [Toronto] Mar 31 p.1.
Augier, M. 1842. *Du Crédit Public.* Paris.
Aumann, Robert J. 2005. "War and Peace – Nobel Lecture 8 December", in Torsten Persson, *Nobel Lectures in Economic Sciences.* Singapore: World Scientific Publishing [in press].
Avery, Simon. 2006. "Data mining: Deeper, smarter and way, way faster; SAS boss says hardware developments are driving advances in business intelligence", *The Globe and Mail Report on Business* [Toronto] May 04, p.8.

Bahree, Bhushan and Jeffrey Ball. 2005a. "Oil Giants Face New Competition For Future Supply; Big Players Focus on Returns As Rivals Undercut Them; Limping Away From Libya" *Wall Street Journal* [NewYork] Apr 19 p.A1.

―――. 2005b. "Saudis Vow Action on Oil Demand, Doubling Spending On Energy Development to US$50-Billion, Will Ignore Output Caps" *Wall Street Journal* [New York] Apr 22 p.A2.

――― and Thaddeus Herrick. 2005c. "Oil Industry's Refining Squeeze Limits Prospects of Price Relief" *Wall Street Journal* [New York] May 24 p.A1.

―――. 2005d. "Big Thirst for Oil Is Unslaked", *Wall Street Journal* [New York] Jun 21 p.A2.

――― and Russell Gold. 2005e. "Pursuit of New Oil Supplies Runs Into a Bottleneck", *Wall Street Journal* [New York] Jun 28 p.A1.

――― and Thaddeus Herrick.2005f_ "Exxon, Aramco Join Sinopec In Refinery Venture in China", *Wall Street Journal* [New York] Jul 11 p.A2.

―――. 2006a. "OPEC Ministers Agree To Keep Oil Output Steady", *Wall Street Journal* [New York] Jan 31 p.A2.

――― and Chip Cummins. 2006b. "Thwarted Attack At Saudi Facility Stirs Energy Fears", *Wall Street Journal* [New York] Feb 25 p.A1.

―――. 2006d. "OPEC Will Keep Oil Output Steady As US Stores Rise", *Wall Street Journal* [New York] Mar 09 p.A2.

―――. 2006e. "Venezuela Seizes Total, ENI Oil Fields", *Wall Street Journal* [New York] Apr 04 pA2.

――― and Ann Davis. 2006f. "Oil Settles Above $70 a Barrel, Despite Inventories at 8-Year High", *Wall Street Journal* [New York] Apr 18 p.A1.

―――, Carla Anne Robbins and Chip Cummins. 2006g. "Oil Minister Iran Won't Cut Exports Despite Nuclear Standoff", *The Wall Street Journal* [New York] Apr 26 p.A1.

Bains, H.S. 1967. *Necessity for Change*. London: The Internationalists.

―――. 1982. *The Necessity for Revolution*. Toronto: MELS Institute.

―――. 2005. *Thinking of the Sixties*. Toronto: National Publications Centre.

Ball, Jeffrey and Chip Cummins. 2005a. "Exxon and Shell Profits Surge Even as Oil Production Declines", *Wall Street Journal* [New York] Apr 29 p.A2.

―――. 2005b. "Exxon Chief Makes A Cold Calculation On Global Warming", *Wall Street Journal* [New York] Jun 14 p.A1.

―――. 2005c. "Texas Truckers Turn To Newfangled Fuel- A Willie Nelson Brand", *Wall Street Journal* [New York] Jul 05 p.A1.

――― and Benoît Faucon. 2005d_ "Exxon, Shell Profits Climb Sharply", *Wall Street Journal* [New York] Jul 29 p.A3.

——, John J. Fialka and Russell Gold. 2005e. "Texas Backlash Spreads As Profits Surge At Oil Companies", *Wall Street Journal* [New York] Oct 28 p.A1.

——, John J. Fialka and Russell Gold. 2005f. "Oil Patch Faces Rough Patch", *Wall Street Journal* [New York] Nov 04 p.A4.

——. 2005g. "US Oil Firms Reach Deal With Libya", *Wall Street Journal* [New York] Dec 30 p.A3.

——. 2006a. "As Exxon Pursues African Oil, Charity Becomes Political Issue", *Wall Street Journal* [New York] Jan 10 p.A1.

——. 2006b. "The New Act at Exxon- Diplomacy of Controlling Future Oil and Gas Reserves", *Wall Street Journal* [New York] Mar 08 p.B1.

——. 2006c. "Energy- Mixed Messages from Big Oil", *Wall Street Journal* [New York] Mar 11 p.A2.

——. 2006d. "With a Big Nuclear Push, France Transforms Its Energy Equation", *Wall Street Journal* [New York] Mar 28 p.A1.

——. 2006e. "As the Price of Gasoline Takes Off, Oil and Auto Firms Trade Barbs", *Wall Street Journal* [New York] Apr 12 p.A1.

——. 2006f. "Exxon's $8.4 Billion Net Faces Fire on All Fronts", *Wall Street Journal* [New York] Apr 28 p.A2.

——. 2006g. "As Gasoline Prices Soar, Americans Resist Major Cuts in Consumption", *Wall Street Journal* [New York] May 01 p.A1.

Baran, Paul A. 1957. *The Political Economy of Growth*. New York: Monthly Review.

——. and Paul M. Sweezy. 1965. *Monopoly Capital* New York: Monthly Review.

Barta, Patrick and Matt Pottinger. 2005a. "Why Cnooc May Not Be Such a Big Threat", *The Wall Street Journal* [New York] Jun 03 p.A1.

—— and Sarah Nassauer .2005b. "Turkey in the Tank- High Price of Gasoline Is a Boon for Biofuels", *The Wall Street Journal* [New York] Oct 28 p.A1.

Basu, A., J. Akhter, M.H. Rahman, and M.R. Islam. 2005. "A review of separation of gases using membranes", *Journal of Petroleum Science and Technology* [in press].

——, Mustafiz, S., N. Bjorndalen, S. Rahaman and M.R. Islam. 2006. "A Comprehensive Approach for Modeling Sorption of Lead and Cobalt Ions through Fish-Scales as an Adsorbent", *Chem. Eng. Comm.*, Vol. 193: 580-605.

Begg, H. M. And S. McDowell.1981. *Industrial Performance and Prospects in Areas Affected by Oil Development*, Edinburgh: Scottish Economic Planning Department, ESU Research Paper No. 3.

Berkeley, George. 1734. "The Analyst", in George Sampson ed., *The Works of George Berkeley* London: G. Bell and Sons, 1898.

Berman, Dennis K. 2005a. "US Seems Wary Of Giving Cnooc Fast Review of Bid", *The Wall Street Journal*[New York] Jun 28 p.A2.

—— and Russell Gold. 2005c. "As Rancor Mounts, Cnooc Needs to Push Its Offer for Unocal", *The Wall Street Journal*[New York] Jul 05 p.C1.

—— and Mark Heinzl. 2005d. "Canada Welcomes China's Cash", *The Wall Street Journal* [New York] Jul 15 p. C1.

—— and Russell Gold. 2005e. "Chevron Raises Its Bid for Unocal", *The Wall Street Journal*[New York] Jul 20 p.A3.

Bilby, Kenneth 1986. *The General: David Sarnoff and the rise of the communications industry* New York:Harper and Row.

Bilmes, Linda and Joseph E. Stiglitz. 2006. "The Economic Costs of the Iraq War: An Appraisal Three Years After the Beginning of the Conflict" Boston MA: American Economics Association Meeting, 8 January.

Blair, Tony. (1998). *The Third Way: New Politics for the New Century*. London: Fabian Society, ISBN 0716305887.

Böhm-Bawerk, Eugen. 1898. *Karl Marx and the Close of His System*. Translated by Alice McDonald. London: T. Fisher Unwin. Reprinted in *Karl Marx and the Close of His System*. New York: Augustus M. Kelley, 1949.

Boyd-Orr, John.1937. *Food, Nutrition and Income*. London: Macmillan.

—— and David Lubbock. 1940. *Feeding the People in Wartime*. London: Macmillan.

——. 1943. *Food and the People*. London: Macmillan.

Bois, C. 1982. "World Hydrocarbon Reserves, Resources and Availabilities", *Revue de l'IFP* Vol 37 No 2. 135:148.

Boyer, Richard O. and Herbert M. Morais 1973. *Labor's Untold Story* New York, NY: UE.

Braverman, Harry 1974. *Labor and Monopoly Capital: The Degradation of Work in the Twentieth Century* New York: Monthly Review Press.

Bradley, Susan and Steve Proctor. 2005. "Goldboro To Get LNG Plant; Eastern Shore To Get $5-Billion Petrochemical Refinery, Says Keltic President", *The Chronicle-Herald* [Halifax] Dec 20 p.A1.

Bradmans. 2005. *Middle East/Africa 2005*. London: Bradmans [periodical; CIP catalogued at British Library].

Brattain, Walter H. 1956. "Surface Properties of Semiconductors – Nobel Lecture 11 December", in *Nobel Lectures, Physics 1942-1962*. Amsterdam: Elsevier [1964].

Brecht, B. and Kurt Weill. 1928. *Der Dreigroschenoper* [The Threepenny Opera].Berlin: Theater am Schiffbauerdamm.

———. 1947. *Selected Poems*. New York: Harcourt-Brace. Translations by H.R. Hays.

Brethour, Patrick R. 2005a. "Following the boom, a bitter divide", *The Globe and Mail Report on Business* [Toronto] Apr 06 pp. 8-9.

———. 2005b. "How to grind the oil sands into a 'sausage factory'", *The Globe and Mail Report on Business* [Toronto] May 21 p.3.

———. 2005c. "North-South energy links grow ever tighter", *The Globe and Mail Report on Business* [Toronto] May 21 p.18.

———. 2005d. "Canadian oil could be headed to China under latest deal", *The Globe and Mail Report on Business* [Toronto] Jun 01p1.

———. 2005e. "Oil soars to record high after terrorist threat, refinery woes", *The Globe and Mail Report on Business* [Toronto] Jun 18 p.5.

———. 2005f. "Petrocan results", *The Globe and Mail Report on Business* [Toronto] Jul 27 p.1.

———. 2005g. "Rig costs cap profits in oil patch", *The Globe and Mail Report on Business* [Toronto] Jul 29 p.1.

——— and Dave Ebner.2005h. "Bidding war may erupt in Alberta oil sands", *The Globe and Mail Report on Business* [Toronto] Aug 04 p.14.

———. 2005i. "Shell doubles oil sands expansion budget", *The Globe and Mail Report on Business* [Toronto] Aug 10 p.1.

———. 2005j. "Why Canada's feeling the pain at the pumps", *The Globe and Mail Report on Business* [Toronto] Sep 02 p.1.

———. 2005k. "Pump prices in Nfld to jump", *The Globe and Mail Report on Business* [Toronto] Sep 03 p.24.

———. 2005l. "Alberta boosts oil exports to aid US", *The Globe and Mail Report on Business* [Toronto] Sep 05 p.1.

———. 2005m. "Oil patch unions seize initiative in tackling labour shortage", *The Globe and Mail Report on Business* [Toronto] Sep 09 p.3.

———. 2005n. "Nexen, OPTI plan $10-billion expansion for oil sands venture", *The Globe and Mail Report on Business* [Toronto] Sep 15 p.1.

———. 2005o. "Fox lends support to Canada over NAFTA", *The Globe and Mail* [Toronto] Sep 30 p.A8.

———. 2005p. "There's still lots of oil – at a price- IEA", *The Globe and Mail Report on Business* [Toronto] Nov 08 p.11.

———. 2005q. "Shell spending to rocket 60% next year", *The Globe and Mail Report on Business* [Toronto] Nov 18 p.3.

———. 2005r. "Pembina Institute for Sustainable Development urges slowdown on oil sands", *The Globe and Mail Report on Business* [Toronto] Nov 24 p.6.

———. 2005s. "Petro-Canada wants stake in Syrian asset sold by year's end", *The Globe and Mail Report on Business* [Toronto] Nov 26 p.3.

———, Bill Curry and Jane Armstrong. 2005t. "Newfoundland takes aim at oil firms", *The Globe and Mail Report on Business* [Toronto] Nov 30 p.10.

———. 2005u. "Oil- A new continental divide", *The Globe and Mail Report on Business* [Toronto] Dec 20 p.1.

———. 2006a. "Husky aims for third offshore project", *The Globe and Mail Report on Business* [Toronto] Jan 17 p.3.

———. 2006b. "How a Tory win could open door to Petrocan sale", *The Globe and Mail Report on Business* [Toronto] Jan 19 p.1.

———. 2006c. "Petrocan to build oil sands upgrader near Edmonton", *The Globe and Mail Report on Business* [Toronto] Jan 27 p.7.

——— and Dave Ebner. 2006d. "Petrocan gets LNG link with Gazprom", *The Globe and Mail Report on Business* [Toronto] Mar 15 p.1.

——— and Sinclair Stewart. 2006e. "How the oil patch is cashing in" *The Globe and Mail Report on Business* [Toronto] Apr 08 p.4.

———. 2006f. "Williams wants expropriation tools", *The Globe and Mail Report on Business* [Toronto] Apr 11 p.1.

——— and Steven Chase. 2006g. "Chevron rushes to disband", *The Globe and Mail Report on Business* [Toronto] Apr 12 p.1.

Brym, Robert J. and R. James Sacouman, edd. 1979. *Underdevelopment and Social Movements in Atlantic Canada* Toronto: University of Toronto – New Hogtown.

Buchanan, Susan. 2006. "Raw-Sugar Futures Fall Off High As Brazil Ethanol Rumors Swirl", *The Wall Street Journal* [New York] Jan 04 p.C2.

Butterfield, Herbert. 1931. *The Whig Interpretation of History.* London: Bell.

———. 1968. *The Origins of Modern Science 1300-1800* Toronto: Clarke, Irwin,.

Byron, R.1986. *Sea Change A Shetland Society, 1970-1979.* St. John's NL: Institute of Social and Economic Research, Memorial University of Newfoundland.

Calmes, Jackie and Guy Chazan. 2005. "Evans Declines Putin's Job Offer At State Oil Firm", *The Wall Street Journal* [New York] Dec 20 p.B11.

Canada. 1976. *Community and Employment Implications of Restructuring the Atlantic Groundfisheries.* Ottawa: Department of Regional Economic Expansion.

―――― . 1983. *Venture Development Project: Report of the Sable Island Environmental Assessment Panel* Ottawa.

―――― . 1999. *Interdepartmental Review of the Canadian Patrol Frigate Project: Report on the Contract Management Framework.* Ottawa: Departments of National Defence, and Public Works and Government Services, 26 March.

―――― . 2004. *Canadian Natural Gas - Review of 2003 and Outlook to 2020.* Ottawa: Department of Natural Resources Energy Policy Sector, Petroleum Resources Branch - Natural Gas Division [December].

Canadian Association of Petroleum Producers [CAPP]. 2006. "Canada's Upstream Oil And Gas Industry Energized- Climate Change Report", *The Globe and Mail Special Report* [Toronto] Feb 23 p.F2.

Campbell, Murray. 2005. "The smokestacks are going cold", *The Globe and Mail* [Toronto] Jul 05 p.A7.

Canning, Stratford G. and C. M. Campbell. 1982. "Prospects for Co-Existence: An Analysis of Potential Interactions between Onshore Petroleum Development and the Established Fishery in Southeastern Newfoundland" NORDCO Ltd. – Paper presented to International Conference on Oil and the Environment, Halifax.

Carey, Alex. 1995. *Taking the Risk out of Democracy: Propaganda in the U.S. and Australia.* Sydney: University of New South Wales Press.

Carlton, Jim. 2006. "BP Finds New Pipeline Rupture Caused By Corrosion In Alaska", *The Wall Street Journal* [New York] Apr 17 p.A3.

Carlyle, Tamsin. 2005a. "Boom in Alberta Oil Sands Fuels Pipeline Dreams to West Coast, China Markets", *The Wall Street Journal* [New York] May 31 p.A2.

―――― . 2005b. "Nexen Targets Canada Oil Sands, To Sell Some Conventional Fields", *The Wall Street Journal* [New York] Jul 06 p.B10.

―――― . 2005c. "Ontario Nuclear Power Plant Units to Be Restarted", *The Wall Street Journal* [New York] Oct 18 p.A11.

―――― . 2006a. "Alberta's Draw- Oil Sands, and Technology", *The Wall Street Journal* [New York] Feb 14 p.A17.

―――― . 2006b. "EnCana Slashes Capital Budget; Profit Falls 8.3%" *The Wall Street Journal* [New York] Feb 16 pA2.

Catton, William R. Jr. 1980. *Overshoot: The Ecological Basis of Revolutionary Change* Chicago: University of Illinois Press.

Champion, Marc. 2006. "Oil-Price Shock Tops List Of Global Economic Risks Amid Supply, Geopolitical Worries", *The Wall Street Journal* [New York] Jan 30 p.A2.

Chase, Steven and Simon Tuck. 2005a. "'National security' bill not aimed at energy takeovers- David Emerson", *The Globe and Mail Report on Business* [Toronto] Jul 15 p.1.

——— and Simon Tuck. 2005b. "EnCana's 'field of dreams' has Colorado locals crying the blues", *The Globe and Mail Report on Business* [Toronto] Jul 15 p.4.

———. 2005c. "Ottawa strikes oil and gas bonanza", *The Globe and Mail* [Toronto] Sep 24 p.A11.

——— and Dave Ebner. 2006a. "Plug pulled on Hebron offshore project", *The Globe and Mail Report on Business* [Toronto] Apr 04 p.1.

——— 2006b. "Eastern businesses seek slice of booming Alberta's oil wealth", *The Globe and Mail Report on Business* [Toronto] Apr 10 p.1.

Chassany, Anne-Sylvaine. 2006. "Total's Net Profit Declines 37% Despite Jump In Energy Prices", *The Wall Street Journal* [New York] Feb 16 p.A2.

Chazan, Guy. 2006. "Russia Says It Plans To Loosen State Monopoly On Gas Exports", *The Wall Street Journal* [New York] Feb 13 p.A4.

Chernenko, K.U. and M S Smirtyukov, edd.1967. Решения партий и правителства по хозяиственным вопросам [*Resheniya partii i pravitel'stva po khozyaystvennym voprosam* – Resolutions of the Party and the Government on Economic Questions], Vol. 1: 1917 – 1928. Moscow.

Chester, Lewis, Stephen Fay and Hugo Young. 1967. *The Zinoviev Letter.* London: Heinemann.

Chhetri, A.B and Islam, M.R. 2006a. "Towards producing a true green biodiesel", *Energy Sources, submitted.*

Chown, Marcus. 2001. "*Principia Mathematica* III – Opinion/Interview with Stephen Wolfram", in *New Scientist* 25 August 2001.

Cohen, A.1980. "A Sense of Time, a Sense of Place: The Meaning of Close Social Association in Whalsay, Shetland", in A. Cohen ed., *Belonging: Identity and Social Organization in British Rural Cultures.* Manchester UK: Manchester University Press and St. John's NL: Institute of Social and Economic Research, Memorial University of Newfoundland.

Cohen, Adam. 2005. "Spain Wins Final Antitrust Say On Energy Deal", *The Wall Street Journal* [New York] Nov 15 p, A20.

Cole, G.D.H. 1944. *A Century of Co-operation.* London: G. Allen and Unwin Ltd.

———. 1956. *A History of Socialist Thought.* London: Macmillan.

Comte, Auguste. 1848. *A General View Of Positivism.* Paris.

Cook, Fred J. 1962. *The Warfare State.* New York: Macmillan.

Cordahi, James and Will Kennedy. 2006. "Saudi Arabia Looks To Eastern Markets", *The Globe And Mail Report On Business* [Toronto] Jan 23 p.7.

Corey, Lewis 1930. *The House of Morgan* New York: Grosset and Dunlap.

Creel, George.1920. *How We Advertised America*. New York: Harper and Brothers.

Cummins, Chip and Thaddeus Herrick. 2005a. "An Oil Giant Faces Questions About a Deadly Blast in Texas", *The Wall Street Journal* [New York] Jul 27 p.A1.

——— . 2005b. "Istanbul Moment- Sir, There's a Ship In Your Bedroom", *The Wall Street Journal* [New York] Jul 28 p.A1.

———, Bhushan Bahree and Jeffrey Ball. 2005c. "Why the World Is One Storm Away From Energy Crisis", *The Wall Street Journal* [New York] Sep 24 p.A1.

——— and Bhushan Bahree, Shai Oster and John Fialka. 2005d. "Five Who Laid the Groundwork For Historic Spike in Oil Market", *The Wall Street Journal* [New York] Dec 20 p.A1.

——— . 2006. "As Oil Supplies Are Stretched, Rebels, Terrorists Get New Clout", *The Wall Street Journal* [New York] Apr 10 p.A1.

Dales, H. Garth and W. Hugh Woodin.1996. *Super-Real Fields*. Clarendon Press.

Darwin, Charles. 1859. *The Origin of Species*. London.

——— . 1871. *The Descent of Man and Selection in Relation to Sex*. London: John Murray.

Darwin, Francis, ed. 1892. *The Autobiography of Charles Darwin and Selected Letters* New York: Dover.

Decloet, Derek.2006. "Oil", *The Globe and Mail* [Toronto] Apr 20 p.A1.

de Córdoba, José. 2005. "Bolivia Election Portends Foreign-Investor Clash", *The Wall Street Journal* [New York] Dec 20 p.A13.

——— . 2006. "In Bolivia, A New Sheriff's In Town", *The Wall Street Journal* [New York] Feb 03 p.A10.

de Pian, Louis 1962. *Linear Active Network Theory* Englewood Cliffs, N.J:. Prentice-Hall.

Deveau, Scott. 2005. "US threatens Pakistan over Iran pipeline plan", *The Globe and Mail* [Toronto] Jun 17 p.S3.

Dowlee, Andrew and Beth Heinsohn. 2005. "More Outages Hit US Oil Refineries", *The Wall Street Journal* [New York] Aug 09 p.A5.

Dummett, Ben. 2005. "Chinese Firms to Pay $1.42 Billion For EnCana Oil Assets in Ecuador", *The Wall Street Journal* [New York] Sep 14 p.A3.

Durkheim, Emil.1897. *Suicide* Paris.

Eagle, Lyon, Pope Associates.1981. "Shetland Offshore Hazard Risk Assessment" Prepared for the Shetland Islands Council.

Easterbrook, W.T. and Hugh G.J. Aitken 1956 *Canadian Economic History* Toronto: Macmillan.

Easterbrook, W.T. and M.H. Watkins, edd. 1967. *Approaches to Canadian Economic History* Toronto:McClelland and Stewart.

Ebner, Dave. 2005a. "High oil prices revive shelved Nfld Project", *The Globe and Mail Report on Business* [Toronto] Apr 06 p.3.

———. 2005b. "The Pembina deposit, Highpine Oil and Gas, the Stollery clan", *The Globe and Mail Report on Business* [Toronto] Apr 07 p.1.

———. 2005c. "Kazakh courts seize Turgai", *The Globe and Mail Report on Business* [Toronto] Apr 16 p.5.

———. 2005d. "Imperial, Exxon pumping up to $6.5-billion more into oil sands", *The Globe and Mail Report on Business* [Toronto] Jun 13 p.1.

———. 2005e. "EnCana unveils plans to sell Alberta hub for gas storage" *The Globe and Mail Report on Business* [Toronto] Jun 21 p.1.

———. 2005f. "How's this for contrarian- World now facing oil glut", *The Globe and Mail Report on Business* [Toronto] Jun 22 p.1.

———. 2005g. "Imperial Oil gets 3rd chief for Mackenzie Valley pipeline project", *The Globe and Mail Report on Business* [Toronto] Jun 25 p.6.

———. 2005h. "PetroKaz soars as company considers sale offers from India, China", *The Globe and Mail Report on Business* [Toronto] Jun 28 p.1.

———. 2005i. "More US oil exploration urged", *The Globe and Mail Report on Business* [Toronto] Jun 29 p.8.

———. 2005j. "Moving fuel -- history of Canadian long-distance pipeline construction", *The Globe and Mail* [Toronto] Jun 30 p.A13.

———. 2005k. "Following China's interest, Snow visits oil sands [!]", *The Globe and Mail Report on Business* [Toronto] Jul 08 p.5.

———. 2005l. "Ottawa, Deh Cho reach pipeline deal", *The Globe and Mail* [Toronto] *Report on Business* Jul 12 p1).

———, Dawn Walton and Deborah Yedlin. 2005m. "Oil-field success leads to Stampede excess", *The Globe and Mail* [Toronto] Jul 13 p.A1.

———. 2005n. "Behind EnCana's swagger, it remains steady as she goes", *The Globe and Mail Report on Business* [Toronto] Jul 14 p.4.

———. 2005o. "Upstart firm gets regulatory nod for plan to sell raw bitumen in oil sands for others to upgrade", *The Globe and Mail Report on Business* [Toronto] Jul 21 p.6.

———. 2005p. "Shell profit gushes 85% to a record on high prices", *The Globe and Mail Report on Business* [Toronto] Jul 22 p.3.

———. 2005q. "EnCana to drill new offshore well near Deep Panuke", *The Globe and Mail Report on Business* [Toronto] Jul 23 p.8.

―――― . 2005r. "French firm snags oil sands project", *The Globe and Mail Report on Business* [Toronto] Aug 03 p.1.

―――― . 2005s. "If $65-a-barrel oil doesn't curb demand, future price shocks will, analyst says", *The Globe and Mail Report on Business* [Toronto]Aug 11 p.2.

―――― . 2005t. "Enbridge inks deals on oil sands pipeline", *The Globe and Mail Report on Business* [Toronto] Sep 10 p.7.

―――― . 2005u. "Shell prepared to spend billions on oil sands in Peace River area", *The Globe and Mail Report on Business* [Toronto] Sep 14 p.3.

―――― . 2005v. "Oil sands trigger race for diluent supply", *The Globe and Mail Report on Business* [Toronto] Sep 21 p.3.

―――― . 2005w. "Oil sands worth $1.4-trillion, study finds", *The Globe and Mail Report on Business* [Toronto] Sep 30 p.1.

―――― . 2005x. "Morgan hands over reins, signals EnCana not for sale", *The Globe and Mail Report on Business* [Toronto]Oct 26 p.1.

―――― . 2005y. "Day after Morgan announces step-down, hedge contracts blast EnCana profit", *The Globe and Mail Report on Business* [Toronto] Oct 27 p.1.

―――― . 2005z. "Elite club of Alberta oil barons sits on billions in paper riches", *The Globe and Mail Report on Business* [Toronto] Oct 28 p.1.

―――― . 2005aa. "CNQ bets $25-billion on oil sands", *The Globe and Mail Report on Business* [Toronto] Nov 03 p.1.

―――― . 2005bb. "EnCana outlines bold expansion in the oil sands", *The Globe and Mail Report on Business* [Toronto] Nov 08 p.1.

―――― . 2005cc. "Saudi oil- Ample or apocalyptically low?", *The Globe and Mail Report on Business* [Toronto] Nov 15 p.9.

―――― and Simon Tuck. 2005dd. "Ottawa set to give Imperial concessions", *The Globe and Mail Report on Business* [Toronto] Nov 18 p.1.

―――― . 2005ee. "Imperial Oil to reveal Mackenzie plans" *The Globe and Mail Report on Business* [Toronto] Nov 23 p.7.

―――― . 2005ff. "Deep Panuke probably will be a go, Hamm says", *The Globe and Mail Report on Business* [Toronto] Nov 24 p.2.

―――― . 2005gg. "Imperial's pipeline plan back on track", *The Globe and Mail Report on Business* [Toronto] Nov 24 .p3.

―――― . 2005hh. "Alberta poised for huge drilling rights sale", *The Globe and Mail Report on Business* [Toronto] Dec 12 p.1.

―――― . 2006a. "Exploration licenses expire offshore Nova Scotia", *The Globe and Mail Report on Business* [Toronto]Jan 04 p.8.

―――― and Dawn Walton. 2006b. "Alberta rejects sour gas wells request", *The Globe and Mail Report on Business* [Toronto] Jan 05 p.1.

――――. 2006c. "EnCana to pay $250,000 for environmental offence", *The Globe and Mail Report on Business* [Toronto] Jan 12 p.18.

――――. 2006d. "New government won't scrap pipeline fund, Imperial says" *The Globe and Mail Report on Business* [Toronto] Jan 16 p.3.

――――. 2006e. "Mackenzie Valley Where Pipe Dreams, Fears Collide", *The Globe and Mail Report on Business* [Toronto] Jan 23 p.1.

――――. 2006f. "India not eyeing oil sands, analyst says", *The Globe and Mail Report on Business* [Toronto] Jan 31 p.7.

――――. 2006g. "TransCanada's Keystone line to ship oil sands crude", *The Globe and Mail Report on Business* [Toronto]Feb 01 p.5.

――――. 2006h. "New battle of Alberta- pipelines", *The Globe and Mail Report on Business* [Toronto] Feb 03 p.3.

――――. 2006i. "Natural gas price crunch seen", *The Globe and Mail Report on Business* [Toronto] Mar 28 p1.

――――. 2006j. "Penn West nabs Petrofund for $3.1-billion", *The Globe and Mail Report on Business* [Toronto] Apr 18 p.1.

――――. 2006k. "Focus Trust buys Profico for $1.1-billion", *The Globe and Mail Report on Business* [Toronto]Apr 25 p.1.

――――. 2006l. "EnCana CEO may spin out oil sands unit", *The Globe and Mail Report on Business* [Toronto] Apr 27 p.1.

EDITORIAL. 2005. "Anti-poverty's moon landing", *The Sunday Herald* [Halifax], Dec 11 p.A16.

El-Rashidi, Yasmine. 2005. "In Kuwait, Gush Of Oil Wealth Dulls Economic Change", *The Wall Street Journal* [New York] Nov 04 p.A1.

Engels, Frederick. 189x. *The Origin of the Family, Private Property, and the State* New York, NY: Pathfinder - 1983 ed.

Faucon, Benoît. 2005. "SEC Asks Oil Firms - Especially Major Non-US Competitors - to Disclose Any Commissions Paid in Iran", *The Wall Street Journal* [New York] May 05 p.A2.

"Feed The Goat", *The Globe and Mail Report on Business*, 13 July 2004, p2.

Feynman, Richard P. 1985. *Surely You're Joking, Mr. Feynman!Adventures of a Curious Character* New York, London: W.W. Norton and Company.

――――. 1988. *What Do You Care What Other People Think?* New York: London, W.W. Norton and Company.

Fialka, John J. 2005a. "Oil, Coal Lobbyists Mount Attack On Senate Plan To Curb Emissions", *The Wall Street Journal* [New York] Jun 21 p.A4.

―, Russell Gold and Rafael Gerena-Morales. 2005b. "US Releases Oil From Stockpile To Ease Crunch", *The Wall Street Journal* [New York] Sep 01p.A3.

―― and Bhushan Bahree. 2005c. "Congress Weighs Oil-Patch Aid, Military's Role", *The Wall Street Journal* [New York] Sep 27 p.A3.

――. 2005d. "Deal Is Near On Offshore Drilling", *The Wall Street Journal* [New York] Oct 06 p.A3.

――. 2005e. "Oil Executives Could Face Probe Of Their Testimony To Congress", *The Wall Street Journal* [New York] Nov 17 p.A10.

――. 2005f. "New England Braces For Energy Squeeze", *The Wall Street Journal* [New York] Dec 21 p.A4.

――. 2006a. "Coalition Turns On To 'Plug-In Hybrids' ", *The Wall Street Journal* [New York] Jan 25 p.A4.

――. 2006c. "Interior Department Seeks New Offshore Leasing", *The Wall Street Journal* [New York] Feb 09 p.A4.

―― and Jeffrey Ball. 2006b. "Bush's Latest Energy Solution, Like Its Forebears, Faces Hurdles", *The Wall Street Journal* [New York] Feb 02 p.A1.

――. 2006d. "Senators Push For Drilling In Gulf Of Mexico", *The Wall Street Journal* [New York] Feb 17 p.A6.

―― and Chip Cummins. 2006e. "Oil-Firm Merger, Tactic Controls Appear To Advance In The Senate", *The Wall Street Journal* [New York] Mar 15 p.A8.

――. 2006f. "Wildcat Producer Sparks Oil Boom On Montana Plains", *The Wall Street Journal* [New York] Apr 05 p.A1.

―, Russell Gold and Laura Meckler. 2006g. "Bush to Seek Overhaul Of Cars' Fuel-Economy Levels", *The Wall Street Journal* [New York] Apr 28 p.A2.

―― and Laura Meckler. 2006h. "Republicans Plan Series of Votes To Gain Control of Energy Issue", *The Wall Street Journal* [New York] May 11 p.A6.

Fisheries And Offshore Oil Consultative Group FOOCG.1983. "Abandonment of Disused Pipelines" Paper presented at a meeting of the FOOCG subgroup on Pipelines at Aberdeen, October.

Fleay, B.J. 1998. "Climaxing oil: How will transport adapt?", *Chartered Institute of Transport in Australia National Symposium - Proceedings* (Launceston).

Frazer, Sir James George. 1890. *The New Golden Bough* Updated ed. by Theodor H.Gaster published 1964 at NY: Mentor Books.

Freeman, Alan. 2005. "Chrétien uses PMO number for business", *The Globe and Mail* [Toronto] Oct 26 p.A7.

Friedman, Milton J. 1948. "A Monetary and Fiscal Framework for Economic Stability", *American Economic Review*, Vol. 38 3., p.245-64. Reprinted in Friedman, 1953.

—— . 1951a. "Some Comments on the Significance of Labor Unions for Economic Policy", in D. McC. Wright, ed., *The Impact of the Union* New York: Harcourt Brace.

—— . 1951b. "Commodity-Reserve Currency", *Journal of Political Economy*, Vol. 59, p.203-32. Reprinted in Friedman, 1953.

—— . 1953. *Essays in Positive Economics*. Chicago: University of Chicago Press.

—— . 1956. "The Quantity Theory of Money: A restatement", in M. Friedman, editor, *Studies in the Quantity Theory of Money*. Chicago: University of Chicago Press. Reprinted in Friedman, 1969.

—— . 1957. *A Theory of the Consumption Function*. Princeton, NJ: Princeton University Press.

—— . 1958. "The Supply of Money and Changes in Prices and Output", in *The Relationship of Prices to Economic Stability and Growth*. Washington, DC: U.S. Congress, Joint Economic Committee. Reprinted in Friedman, 1969.

—— . 1959a. *A Program for Monetary Stability*. New York: Fordham University Press.

—— . 1959b. "The Demand for Money: Some theoretical and empirical results", *Journal of Political Economy*, Vol. 67 4., p.327-51.

—— . 1961. "The Lag in the Effect of Monetary Policy", *Journal of Political Economy*, Vol. 69, p.447-66. Reprinted in Friedman, 1969.

—— . 1962a. *Capitalism and Freedom*. 1977 edition, Chicago: University of Chicago Press.

—— . 1962b. "Should There be an Independent Monetary Authority?", in L.B. Yeager, editor, *In Search of a Monetary Constitution*. Cambridge, Mass: Harvard University Press.

—— . 1963. *Inflation: Causes and consequences*. New York: Asia Publishing House.

—— . 1966a. "Interest Rates and the Demand for Money", *Journal of Law and Economics*, Vol. 9, p.71-85. Reprinted in Friedman, 1969.

—— . 1966b. "What Price Guideposts?", in G.P. Schultz, R.Z. Aliber, editors, *Guildelines: Informal controls and the market place*. Chicago: University of Chicago Press.

—— . 1968a. "The Role of Monetary Policy", *American Economic Review*, Vol. 58, p.1-17. Reprinted in Friedman, 1969.

―――. 1968b. "Money: the Quantity Theory", *International Encyclopedia of the Social Sciences*, p.432-37. Reprinted in Friedman, 1969.

―――. 1969. *The Optimum Quantity of Money and Other Essays*. London: Macmillan.

―――. 1970a. "Comment on Tobin", *Quarterly Journal of Economics*, Vol. 74, p.318-27.

―――. 1970b. "A Theoretical Framework for Monetary Analysis", *Journal of Political Economy*, Vol. 78 2., p.193-238.

―――. 1970c. *The Counter-Revolution in Monetary Theory*. London: Institute of Economic Affairs.

―――. 1971. "A Monetary Theory of National Income", *Journal of Political Economy*, Vol. 79, p.323-37.

―――. 1974a. "Comments on the Critics", in Gordon, 1974.

―――. 1974b. *Monetary Correction: A proposal for escalation clauses to reduce the cost of ending inflation*. London: Institute of Economic Affairs.

―――. 1976a. "Comments on Tobin and Buiter", in J. Stein, editor, *Monetarism*, Amstedam: North-Holland.

―――. 1976b. "Inflation and Unemployment – The Nobel Prize Lecture" 13 December., in Assar Lindbeck, ed., *Nobel Lectures in Economic Sciences 1969-1980* (Stockholm: Stockholm University, 1981.

―――. 1977. "Inflation and Unemployment", *Journal of Political Economy*, Vol. 85 3., p.451-72.

―――. 1984. "Monetary Policy: Tactics versus strategy", in Moore, editor, *To Promote Prosperity*. Stanford, Calif: Hoover Institute.

―――. 1985. "The Case for Overhauling the Federal Reserve", *Challenge*, July/August, p.4-12.

―――. 1987. "Quantity Theory of Money", in J. Eatwell, M. Milgate, P. Newman, editors, *The New Palgrave: A dictionary of economics*. London: Macmillan.

――― and R. Friedman. 1980. *Free to Choose: A personal statement*. New York: Harcourt Brace Jovanovich.

――― and D. Meiselman. 1963. "The Relative Stability of Monetary Velocity and the Investment Multiplier in the United States, 1898-1958", in Commission on Money and Credit, *Stabilization Policies*. Englewood Cliffs, NJ: Prentice-Hall.

――― and A.J. Schwartz. 1963. *A Monetary History of the United States, 1867-1960*. 1971 edition, Princeton: Princeton University Press.

――― and A.J. Schwartz. 1963. "Money and Business Cycle", *Review of Economics and Statistics*, Vol. 30, p.32-64.

—— and A.J. Schwartz. 1970. *Monetary Statistics of the United States: Sources, methods.* New York: Columbia University Press.

—— and A.J. Schwartz. 1982. *Monetary Trends in the United States and the United Kingdom: Their relations to income, prices and interest rates, 1876-1975.* Chicago: University of Chicago Press.

—— and A.J. Schwartz. 1986. "Has Government Any Role in Money?", *Journal of Monetary Economics*, Vol. 17 1., pp. 37-62.

Fukuyama, Francis. 1992. *The End of History and the Last Man.* New York: Free Press.

Galbraith, John Kenneth.1967. *The New Industrial State* Boston: Houghton-Mifflin.

Galina, Andrea, ed. 2003. *Globalization and Meso-Regions* Denmark: Roskilde University.

Gallegos, Raul. 2005. "Venezuela's Oil Moves Signal a New Policy", *The Wall Street Journal* [New York] Jun 14 p.A12.

Gaskin, M. ed. 1978. *The Economic Impact of North Sea Oil on Scotland* London UK: Her Majesty's Stationary Office.

George, Susan 1979. *How the Other Half Dies: The Reasons for World Hunger* England:Penguin.

Gever, J., Kaufmann, R., D. Skole, and C. Vorosmarty. 1991. *Beyond Oil.* Boulder: University Press of Colorado.

Gillman, L. and M. Jerison. 1960. *Rings of Continuous Functions* Van Nostrand.

Glazebrook, G.P. deT. 1966. *A History of Canadian External Relations* Toronto: McClelland and Stewart. 2 vols.

Godfrey, David and Douglas Parkhill eds. 1980. *Gutenberg Two: The New Electronics and Social Change* Toronto: Press Porcépic.

Gold, Russell. 2005a. "Boom Town – Drilling for Natural Gas Faces A Sizable Hurdle As Fort Worth, The US's Largest Field, Lies Under 1.6 Million People And Not Everyone - Especially Those Holding No Subsurface Rights - Can Benefit", *The Wall Street Journal* [New York] Apr 29 p.A1.

——, Matt Pottinger and Dennis K. Berman. 2005b. "China's CNOOC Lobs In Rival $18.5-Billion Bid To Acquire Unocal, Stop Chevron", *The Wall Street Journal* [New York] Jun 23 p.A1.

—— and Greg Hitt. 2005c. "Chevron Labors to Derail Rival's Unocal Bid", *The Wall Street Journal* [New York] Jun 30 p.A4.

——. 2005d. "In Deal for Unocal, Chevron Gambles On High Oil Prices", *The Wall Street Journal* [New York] Aug 10 p.A1.

────── and Thaddeus L. Herrick. 2005e. "Damage to Oil and Gas Facilities Pushes U.S. Closer to Energy Crisis", *The Wall Street Journal* [New York] Sep 02 p.A1.

────── 2005f. "Big Oil Firms Join Hunt For Natural Gas in US", *The Wall Street Journal* [New York] Nov 29 p.A1.

────── . 2005g. "Bidding War Chills U.S. Plan To Import Gas", *The Wall Street Journal* [New York] Dec 19 p.C1.

────── . 2006. "As Prices Surge, Oil Giants Turn Sludge Into Gold", *The Wall Street Journal* [New York] Mar 27 p.A1.

Goldblatt, Robert. 1998. *Lectures on the hyperreals : an introduction to nonstandard analysis* Springer.

Gongloff, Mark. 2006. "Is 'Dark Matter' in the Deficit? Spackle for Economic Anomalies Looks to Explain How U.S. Operates With Massive Debt", *The Wall Street Journal Online* [New York] Feb 10.

Goulden, Joseph, C. 1968. *Monopoly* New York:Putnam.

Grant, John P. 1978. "The Conflict Between the Fishing and the Oil Industries in the North Sea; A Case Study", *Ocean Management,* 4:137-49. Glasgow: Department of Public International Law, University of Glasgow.

Grare, Frederic and Georges Perkovich. 2006. "Baluchistan", *The Wall Street Journal* [New York] Jan 16 p.A15.

Graveland, Bill. 2005. "Shell seeks to double Athabasca oil sands production" *The Globe and Mail Report on Business* [Toronto] Apr 30 p.2.

Grinnell, H. Rae. 1981. *The Implications of Offshore Petroleum Development for Offshore Atlantic Fisheries: A Social-Economic Overview* Ottawa: Department of Fisheries and Oceans, Economic Research Division.

Gromyko, A.A. *et al.* 1973. *Soviet Peace Efforts on the Eve of World War II* Moscow: Novosti. 2 vols.

Gunder Frank, Andre. 1969. *Capitalism and Underdevelopment in Latin America: Historical studies of Chile and Brazil* New York, London: Monthly Review Press.

────── . 2000. "Immanuel and Me With-Out Hyphen", in *Journal of World Systems Research,* vol XI, no 2 [Summer/Fall Special Issue: Festchrift for Immanuel Wallerstein – Part I] pp. 216-231.

Hackett, Robert A., Richard Gruneau *et al.* 2000. *The Missing News:Filters and Blind Spots in Canada's Press* Ontario:Garamond Press.

Hardy, G.H. 1940. *Ramanujan: Twelve Twelve Lectures on Subjects Suggested by His Life and Work* Cambridge UP.

Harris, Michael. 1998. *Lament for an Ocean: The Collapse of the Atlantic Cod Fishery- A True Crime Story.* Toronto: McClelland and Stewart.

Hasbrouck, Joel, George Sofianos and Deborah Sosebee. 1993. "New York Stock Exchange Systems and Trading Procedures", *New York Stock Exchange Working Paper #93-01*. New York: New York University Stern School of Business.

Hausmann, R. and Federico Sturzenegger.2005. "U.S. and global imbalances: Can 'dark matter' prevent a Big Bang?" in Centre for International Development, *Report* Cambridge MA: Harvard University, 30 November.

Hayashi, Yuka. 2005. "China Studies Japan's Mistakes As the Pursuit for Oil Continues", *The Wall Street Journal* [New York] Aug 03 p.A2.

Hayes, Dennis. 1989. *Behind the Silicon Curtain: The Seductions of Work in a Lonely Era* Boston: South End Press.

Heber, Robert W. 1981. "Social Environmental Impacts of Offshore Petroleum Development in Nova Scotia" M.A. Thesis, Institute for Resource and Environmental Studies, Dalhousie University, Halifax.

Hegel, Georg Wilhelm Friedrich. 1833. *The Philosophy of Right*. Berlin.

Heen, Knut. 1984. *Labour Market Behaviour of Fishermen* Trømso NO; Institute of Fisheries, University of Trømso.

Henig, Robin Marantz. 2006. "Looking for the Lie", in *The Sunday New York Times Magazine* New York: The Times Publishing Co., 5 February.

Henwood, Douglas. 2005. "The long strange career of Jeffrey D. Sachs", *Left Business Observer* (Number 111, August).

Herrick, Thaddeus L. 2005a. "As Oil Tops $60, What's Next", *The Wall Street Journal* [New York] Jun 28 p.C1.

——— . 2005b. "Refiners Upgrade Oil Processing, Not the Capacity", *The Wall Street Journal* [New York] Aug 04 p.A2.

——— , Bhushan Bhahree and Keith Johnson. 2005c. "US Resists Building Refineries As Overseas Firms Move Ahead", *The Wall Street Journal* [New York] Dec 28 p.A2.

Hickman T. Alex 1984. *Royal Commsion Report on the Ocean Ranger Marine Disaster*.

Ottawa: Canada Supply and Services.Hobson, John A. 1902. *Imperialism* London.

Higgins, Andrew. 2005a. "Democracy Project In Bahrain Falters; Gulf Kingdom Reverses Course", *The Wall Street Journal* [New York] May 11 p.A1.

——— . 2005b. "As Oil Riches Gush, a Sheik Loosens His Grip on Economy", *The Wall Street Journal* [New York] Oct 21 p.A1.

——— . 2005c. "Oil-Rich Norway Hires Philosopher As Moral Compass", *The Wall Street Journal* [New York] Dec 01 p.A1.

Highham, Charles 1983. *Trading With The Enemy: The Nazi-American Money Plot 1933-1949*. New York: Delacorte.

Hilferding, Rudolf. 1910. *Das Finanzkapital – Eine Studie über die jüngste Entwicklung des Kapitalismus* [Finance Capital: A study in the latest development of capitalism]. Vienna.

Hill, Robert H.1983. *The Social Context of Work and the Reality of Unemployment in Newfoundland Society.* St. John's NL: Community Services Council of Newfoundland and Labrador.

Hilsenrath, Jon E. 2005a. "Energy Network Is Further Strained", *The Wall Street Journal* [New York] Aug 31 p.A5.

———. 2005b. "Novel Way to Assess School Competition Stirs Academic Row", *The Wall Street Journal* [New York] Oct 24 p.A1.

Hobbes, Thomas. 1651. *The Leviathan* London, 1st ed.

Holland, W. J. 2000. *The Navy*. Washington DC: Naval Historical Foundation.

House, J. D.1985. *The Challenge of Oil: Newfoundland's Quest for Controlled Development.* St. John's NL: Institute of Social and Economic Research, Memorial University of Newfoundland.

———, ed. 1986. *Fish vs Oil: Resources and Rural Development in North Atlantic Societies* St. John's NL: Institute of Social and Economic Research, Social and Economic Papers Series No. 16, Memorial University of Newfoundland.

Howlett Karen. 2005a. "New Brunswick Premier seeks national nuclear plan", *The Globe and Mail* [Toronto] Aug 10 p.A4.

———. 2005b. "Energy rebates withheld- Regulator keeps $570-million it owes electricity users, but money will cushion coming rate" *The Globe and Mail* [Toronto] Nov 17 p.A9.

Hunt, Michael H. 1983. *The Making of a Special Relationship: The United States and China to 1914*. New York: Columbia UP.

Hunter, G. L.1976. "Fisheries and Oil", in John Button ed., *The Shetland Way of Oil* Sandwick, Shetland: Thuleprint Limited.

Immen, Wallace. 2005. "Oil patch turnaround uncorks a gusher of jobs", *The Globe and Mail Report on Business* [Toronto] May 21 p.9.

Inhaber, H., and Saunders, H. 1994. "Road to Nowhere", *The Sciences*, Vol. 34, No. 6, November/December 1994.

Innis, Harold A. 1962. *Essays in Canadian Economic History* Toronto: University of Toronto Press.

———. 1954. *The Cod Fishery* Toronto: U of T Press.

———. 1930. *The Fur Trade in Canada* Toronto: U of T Press.

Ip, Greg. 2005. "Will Katrina Cause The Fed to Pause", *The Wall Street Journal* [New York] Sep 01 p.A2.

Irvine, W. 1955. *Apes, Angels, and Victorians: The Story of Darwin, Huxley, and Evolution.* McGraw-Hill, New York.

Islam, M.R., S. Sevgur and T. Lay. 2002. "An Accurate And Remotely Accessible Lie-Detection Machine" [Draft Research Proposal – from M.R. Islam's files].

Islam, M.R. 2003. *Revolution in Education.* Halifax [Canada]: EEC Research Group.

——— . 2005. "Unraveling The Mysteries Of Chaos And Change: The Knowledge-Based Technology Development", *Proceeding of the First International Conference on Modeling, Simulation and Applied Optimization.* Sharjah, U.A.E.: 1-3 February.

Islam, M.R., 2005b. "Knowledge-Based Technologies For The Information Age", JICEC05-Keynote speech, *Jordan International Chemical Engineering Conference V.* Amman, Jordan: JICEC'05, Sep 12-15.

——— and G.M. Zatzman. 2004a. "A New Energy Pricing Model", *MPC 2004* Tripoli [Libya]: International Energy Foundation, February.

——— and G.M. Zatzman. 2005a. "A New Energy Pricing Model: The escalating failure of neoclassical economic models and some possible implications for future relations between the Dollar and the Euro", *CSCE 2005* Toronto: Canadian Society of Civil Engineers 33rd Annual Meeting, June.

——— and G.M. Zatzman. 2005b. "A New Energy Pricing Model: Exploring the evolving relationship between international reserve currencies and global shifts in access to and control over strategic energy resources", *ICERD-3.* Kuwait: 3rd International Conference on Energy Research and Development, November.

——— and G.M. Zatzman. 2006. "Natural Gas Pricing", in S. Mokhatab, J.G. Speight and and W.A. Poe, edd., *Handbook of Natural Gas Transmission and Processing* (Elsevier).

——— , G.M. Zatzman and R. Shapiro. 2006. "The Energy Crunch: What More Lies Ahead", in M. Masood, *Global Dialogue on Energy.* No. 2 in a series, convened 3-4 April in Washington DC at the Centre for Strategic and International Studies.

Iverson, Noel and D. Ralph Mathews 1968. *Communities in Decline: An Examination of Household Resettlement in Newfoundland* St. John's: Newfoundland Institute of Social and Economic Research.

Jenkins, Holman W., Jr. 2005. "The Real 'Oil Crisis' ", *The Wall Street Journal* [New York] Dec 28 p. A15.

——— . 2006. " 'Le Gazprom' " *The Wall Street Journal* [New York] Mar 15 p.A23.

Jevons, W. Stanley. 1865. *The Coal Question* London.
———. 1870. *Theory of Political Economy* London.
Johnson, Avery. 2005. "Hotel Industry Awakening To Bedbug Problem; With Pesticides Out of Favor, Critters Show Up More; A 1920s Cure = Gasoline", *The Wall Street Journal* [New York] Apr 21 p.A1.
Johnson, H.G. 1971. "The Keynesian Revolution and the Monetarist Counter-Revolution", *American Economic Review*, Vol 2, pp.1-14.
Johnson, Keith.2006. "Spain Looks Set To Clear Gas Natural-Endesa Deal", *The Wall Street Journal* [New York] Feb 02 p.A6.
Johnson, Mark. 2005. "Ethanol burns more than it saves- study", *The Globe and Mail Report on Business* [Toronto] Jul 18 p.6.
Jones, Jeffrey. 2006. "India To Invest $1-Billion In Oil Sands", *The Globe And Mail Report On Business* [Toronto] Feb 01 p.5.
Kaldor, N. 1982. *The Scourge of Monetarism* Oxford: Oxford University Press.
Kato, Hiromi. 2005. "Effects of the Oil Price Upsurge on the World Economy", *Journal of the Institute of Energy Economics Japan* [December].
Keisler, H Jerome.1976. *Elementary Calculus* Boston MA: Prindle, Weber and Schmidt.
Kennedy, Peter. 2005. "Ottawa to review US takeover of Terasen", *The Globe and Mail* [Toronto] Oct 21 p.S1.
Kermisch, Ron and Paul Smith. 2005. "Telecom's Other 'Merger'", *The Wall Street Journal* [New York] May 17 p.C1.
Keynes, J.M. Lord. 1936.The General Theory of Employment, Interest and Money. London: Macmillan Cambridge UP.
Khan, M.I., A.B. Chhetri and M.R. Islam. 2005a. Community-Based Energy Model: A Novel Approach in Developing Sustainable Energy [in press].
———, G.M. Zatzman and M.R. Islam. 2005b. "A Novel Sustainability Criterion as Applied in Developing Technologies and Management Tools", *Jordan International Chemical Engineering Conference V.* Amman, Jordan: JICEC'05, Sep 12-15.
———, A.B. Chhetri and M.R. Islam. 2006. "Analyzing sustainability of community-based energy development technologies", *Energy Sources*: in press.
——— and M.R. Islam. 2006. *Achieving True Sustainability in Technological Development and Natural Resources Management.* New York: Nova Science Publishers [in press].
Khan, M.M. and M.R. Islam. 2004. "Down-hole separation of petroleum fluids", *J. Petroleum Science and Technology* [in press].

———, D. Prior, and M. R. Islam. 2005a. "Zero-waste living with inherently sustainable technologies", *Jordan International Chemical Engineering Conference V.* Amman, Jordan: JICEC'05, Sep 12-15.

———, A. R. Mills, M.Y. Mehedi, O. Chaalal and M.R. Islam. 2005b. "Bioabsorbents for the removal of heavy metals from aqueous streams", *Jordan International Chemical Engineering Conference V.* Amman, Jordan: JICEC'05, Sep 12-15.

———, D. Prior, and M. R. Islam. 2005d. "Direct-usage solar refrigeration: from irreversible thermodynamics to sustainable engineering", *Jordan International Chemical Engineering Conference V.* Amman, Jordan: JICEC'05, Sep 12-15.

———, 2006. Personal communication, Department of Civil and Resource Engineering, Dalhousie University, Halifax, Feb 10.

Khazanie, Ramakant 1976. *Basic Probability Theory and Applications* Pacific Palisades: California, Goodyear, Pacific Palisades, California.

Kilby, Jack. 2000. "Turning Potential Into Realities: The Invention of the Integrated Circuit - The Nobel Lecture" (8 December), in Ekspong, Gösta, ed. 2003. *Nobel Lectures in Physics 1996-2000.* Singapore: World Scientific Publishers. pp.474-485.

King, Neil Jr., Greg Hitt and Jeffrey Ball. 2005. "Oil Battle Sets Showdown Over China", *The Wall Street Journal* [New York] Jun 24 p.A1.

Kirby, Michael J. 1982. *Navigating Troubled Waters: Report of the Task Force on Atlantic Fisheries*, Chairman. Ottawa: Supply and Services Canada.

Klein, Naomi. 2002. *No Logo.*

Kline, Morris 1972. *Mathematical Thought from Ancient to Modern Times* New York: Oxford University Press.

Kondratieff, N. 1935. "The Long Waves in Economic Life", *Review of Economic Statistics* Vol XVII, No 6 - November.

Koring, Paul. 2006. "US 'Addicted To Oil,' Bush Concedes", *The Globe And Mail* [Toronto] Feb 01 p.A1.

Kranhold, Kathryn And John M Biers. 2005. "GE In Talks to Buy Technology For Offshore Oil, Gas Production", *The Wall Street Journal* [New York] Aug 01 p.A2.

Kumar, Himendra. 2005. "India Rejects Plan To Buy Stake In Nigeria Oil Field", *The Wall Street Journal* [New York] Dec 19 p.A15Larkin, John. 2005. "Oil-Well Fire in 'Bombay High' Oil Field Complicates Indian Energy Picture", *The Wall Street Journal* [New York] Jul 28 p.A9.

Kunstler, James Howard. 2004. "The Long Emergency: What's going to happen as we start running out of cheap gas to guzzle?", in *Rolling Stone* [New York], Mar 24.
Kuznets, Simon. 1971. "Modern Economic Growth: Findings and Reflections – The Nobel Prize Lecture" 11 December., in Assar Lindbeck, ed., *Nobel Lectures in Economic Sciences 1969-1980* Stockholm: Stockholm University, 1981.
Lipson, Michael. 1999. "The Reincarnation Of Cocom: Explaining Post-Cold War Export Controls", *The Nonproliferation Review* [Winter] pp. 33-51. Monterey CA: Monterey Institute of International Studies.
Lange, Oskar. 1938. *On the Economic Theory of Socialism*. Minneapolis: University of Minnesota.
Leontieff, W. 1973. "Structure of the World Economy: Outline of a Simple Input-Output Formulation Reflections – The Nobel Prize Lecture" 11 December. in Assar Lindbeck, ed., *Nobel Lectures in Economic Sciences 1969-1980* Stockholm: Stockholm University, 1981.
Lenin, V.I. 1916. *Imperialism, The Highest Stage of Capitalism* Peking: Foreign Language Press – 1975 ed.
Levine, Steve, Christopher Cooper and Michael Corkery. 2005. "Katrina's Oily Wake", *The Wall Street Journal* [New York] Sep 12 p.B1.
——— and Jeffrey Ball. 2006. "Exxon's Reliance On Qatar Field Raises Concerns", *The Wall Street Journal* [New York] Feb 16 p.A2.
Lewis, T.M. And I.H. Mcnicoll. 1978. *North Sea Oil and Scotland's Economic Prospects* London: Croon Helm.
Lienhard, John H. 2000. *The Engines of Our Ingenuity: An Engineer Looks at Technology and Culture* Oxford UP.
Lindbeck, Assar, ed. 1981. *Nobel Lectures in Economic Sciences 1969-1980* Stockholm: Stockholm University.
Linebaugh, Kate, Matt Pottinger, Greg Hitt and Jason Singer. 2005. "After Earlier Fumbles, Cnooc Uses Wall Street Tactics in Unocal Bid", *The Wall Street Journal* [New York] Jun 27 p.A1.
Luhnow, David and Geraldo Samor. 2006a. "As Brazil Fills Up on Ethanol, It Weans Off Energy Imports", *The Wall Street Journal* [New York] Jan 09 pA1.
———. 2006b. "How Brazil Broke Its Oil Habit", *The Wall Street Journal* [New York] Feb 06 pA9.
———. 2006c. "Mexico's Oil Output May Decline Sharply", *The Wall Street Journal* [New York] Feb 09 pA4.

——— and Peter Millard. 2006d. "Chávez Plans to Take More Control Of Oil Away From Foreign Firms", *The Wall Street Journal* [New York] Apr 24 pA1.

——— and José De Córdoba. 2006e. "Bolivia Seizes Natural-Gas Fields In a Show of Energy Nationalism", *The Wall Street Journal* [New York] May 02 pA1.

Lyons, John.2006. "Panama Takes Step Toward Expanding The Canal", *The Wall Street Journal* [New York] Apr 24 p.A8.

Macgillivray, Don and Brian Tennyson 1981. *Cape Breton Historical Essays* Nova Scotia: College of Cape Breton Press.

Mackay, R.A. 1971. *Canadian Foreign Policy 1945-1954: Selected Speeches and Documents* Toronto:McClelland and Stewart.

Macpherson, C.Brough. 1967. *The Real World of Democracy* Toronto: CBC.

———. 1964. *The Political Theory of Possessive Individualism* Oxford: Oxford UP.

Maher, Kris. 2005. "Making Out Like Bandits As Oil Price Hikes Push Coal from $30 to $50 a Ton", *The Wall Street Journal* [New York] May 05 p.A2.

Mahoney, Jill. 2005. "Canada's added girth a growing concern", *The Globe and Mail* [Toronto], Dec 10.

Malthus, Thomas. 1798. *Essay on Population* London.

Marshall, Alfred. 1890. *Principles of Economics* London: Macmillan.

Martinez, Michael J. 2006. "Mixed Data Could Keep Stock Market Murky", *Associated Press* Business Wire, filed Sunday, Feb 05 at 17:01 ET.

Marx, K. 1892. *Capital: A Critique of Political Economy Vol. III* London, Edited by Frederick Engels.

———. 1883. *Capital: A Critique of Political Economy Vol. II* London, Edited by Frederick Engels.

———. 1867. *Capital: A Critique of Political Economy Vol. I* London, English ed translated from the German by Samuel Aveling.

———. 1859. *A Contribution to the Critique of Political Economy*. English ed. Chicago: Charles H. Kerr, 1918.

Massachusetts Institute Of Technology MIT.1973. *The Georges Bank Petroleum Study* Cambridge MA: Offshore Oil Task Group - Vol.1.

McCarthy, Shawn. 2005a. "Oil price seen sparking alternatives", *The Globe and Mail Report on Business* [Toronto] May 21 p.4.

———. 2005b. "IEA turns on the oil reserves taps", *The Globe and Mail Report on Business* [Toronto] Sep 03 p.24.

———. 2006. "Exxon Plays Hardball – And Hebron One Example", *The Globe And Mail Report On Business* [Toronto] Apr 17 p.1.

McChesney, Robert W. 2004. *The Problem of the Media: U.S Communication Politics in the 21st Century* New York: Monthly Review Press.

McCloy, John J., Nathan W. Pearson and Beverley Matthews. 1976. *The Great Oil Spill.* New York: Chelsea House.

McDonald, Joe. 2005. "The man at the heart of CNOOC's quest", *The Globe and Mail Report on Business* [Toronto] Jun 30 p.15.

McKenna, Barrie and Patrick Brethour. 2005a. "Katrina hits refineries; strategic reserves tapped", *The Globe and Mail Report on Business* [Toronto] Sep 01 p.1.

──── . 2005b. "Race for Arctic pipeline heats up for Canada, US", *The Globe and Mail Report on Business* [Toronto] Oct 08 p.6.

──── . 2005c. "Canada- Energy Nation", *The Globe and Mail Report on Business* [Toronto] Dec 15 p.1.

McKinnon, John D, John J Fialka and Jeffrey Ball.2006??. "Bush Takes Steps To Expand Oil Supplies", *The Wall Street Journal* [New York] Apr 26 p.A4.

McKinnon, Mark. 2005. "The great Caspian Sea adventure-bubble", *The Globe and Mail Report on Business* [Toronto] May 24 p.6.

McLean, Catherine. 2005. "Kinder Morgan enters oil sands with $3-billion Terasen purchase", *The Globe and Mail Report on Business* [Toronto] Aug 02 p.1.

McLuhan, H. Marshall. 1964. *Understanding Media: the Extensions of Man* New York: Signet, New American Library.

────. 1969. *The Gutenberg Galaxy* New York: Signet - New American Library.

Mcnicoll, I.H. 1980 "The Impact of Oil on the Shetland Economy" *Managerial and Decision Economics,* 12.

──── . 1982 "Ex-Post Appraisal of an Input-Output Forecast" *Urban Studies,*19:397-404.

──── and G. Walter.1971. *The Shetland Economy 1976-1977* Lerwick: Shetland Islands Council.

McQuaig, Linda. 2004. *It's the Crude, Dude: War, Big Oil, and the Fight for the Planet.* Toronto-New York: Doubleday.

Melzak, Z.A. 1983. *Bypasses: A Simple Approach to Complexity* Toronto:Wiley,.

Menger, Carl. 1871. *Principles of Economics* Vienna.

Mikesell, R.F., William H. Bartsch *et al.*1971. *Foreign Investment in the Petroleum and Mineral Industries* Baltimore MD: John Hopkins Press for Resources for the Future.

Millard, Peter. 2005. "Exxon Resists Venezuelan Contract Overhaul", *The Wall Street Journal* [New York] Dec 20 p.A12.

———. 2006. "Venezuela To Raise Tax On Foreign Firms In Orinoco Oil Field", *The Wall Street Journal* [New York] Mar 15 p.A8.

Mills, C. Wright. 1951. *White Collar: The American Middle Classes*. New York: Oxford UP.

———. 1956. *The Power Elite*. New York: Oxford UP.

Mittelstaedt, Martin. 2005. "Pollution Debate Born Of Chemical Valley's Girl-Baby Boom", *The Globe And Mail* [Toronto] Nov 15 p.A3.

Mobil Oil Canada Limited. 1983. *Venture Development Project Environmental Impact Statement* Halifax. Submitted to the Government of Canada and Province of Nova Scotia.

Moore, Barrington, Jr. 1967. *Social Origins of Dictatorship and Democracy: Lord and Peasant in the Making of the Modern World* Boston: Beacon Press.

Moore, Oliver. 2005. "US House Drops Plans To Drill In Arctic Refuge", *The Globe And Mail* [Toronto] Nov 10 p.A1.

Morgan, Bernice. 1992. *Random Passage*. St. John's: Breakwater Books.

Morgan, Dan, 1980. *Merchants of Grain* New York:Penguin.

Mosselmans, Bert.1999. "Reproduction and Scarcity: the Population Mechanism in Classicism and in the 'Jevonian Revolution'", *The European Journal of the History of Economic Thought* Vol. 6, No. 1 Spring 1999.

Mufson, Steven and Shailagh Murray. 2006. "Profits, Prices Spur Oil Outrage", Washington Post [Washington DC] Apr 28 p.A01.

Mullens, Brody. 2005a. "Senate Overwhelmingly Passes Energy Bill Long Sought by Bush", *The Wall Street Journal* [New York] Jun 29 p.A2.

———. 2005b. "White House Backs Oil Firms' Attempt To Avoid New Tax", *The Wall Street Journal* [New York] Nov 18 p.A6.

Mustafiz, S., A. Basu, A. Dewaidar, O. Chaalal, and M.R. Islam. 2002. "A Novel Method for Heavy Metal Removal from Aqueous Streams", from: S. Mustafiz, MASc thesis, Dalhousie University, Canada.

Myers, Ransom A. and Boris Worm. 2003. "Rapid worldwide depletion of predatory fish communities", *Nature* v.423: 280-283 [May 15].

Myrden, Judy. 2006a. "Marauder to hunt for natural gas off coast", *The Chronicle Herald,* [Halifax] Oct 04 p.C1.

———. 2006b. "Will Sable still be able? Dalhousie professor says time is running out for natural gas project", *The Chronicle-Herald*, [Halifax] Jan 09 p.C1.

———. 2006c. "Offshore may see delays; Industry study suggests LNG supplies could hinder projects", *The Chronicle-Herald*, [Halifax] Feb 09 pC1.

———. 2006d. "Goldboro project gets a lift", *TheChronicleHerald* [Halifax] Mar 21p.C1.

―――. 2006e. "More than just pipe dreams; Region has LNG advantages, but must adapt to needs of rapidly changing industry", *The Chronicle-Herald* [Halifax] Apr 10 p.C1.

Nadesan, S. 1993. *A History of the Up-Country Tamil People in Sri Lanka* Sri Lanka: Ranko Printers and Publishers.

Naylor, Tom. 1975a. *The History of Canadian Business 1867 -1914-Vol I: the Banks and Finance Capital* Totonto: Lorimer.

―――. 1975b. *The History of Canadian Business 1867-1914-Vol II: Industrial Development* Toronto: Lorimer.

Neidorf, Robert 1967. *Deductive Forms : An Elementary Logic* New York: Harper and Row.

New England River Basin Commission Reports NERBC/RALI. 1976. *Onshore Facilities Related to Offshore Oil and Gas Development: Estimates for New England* Boston MA.

Newton, Sir Isaac.1687. *Mathematical Principles of Natural Philosophy* [1729 translation of *Principia Mathematica* from Latin original, by Andrew Motte] London.

Nicolson, J.R. 1975. *Shetland and Oil* London: William Luscombe.

NORDCO. 1981a. *"It Were Well to Live Mainly Off Fish": The Place of the Northern.*

Cod in Newfoundland's Development. St. John's NL.

―――. 1981b. *A Study of the Potential Social-Economic Effects on the Newfoundland Fishing Industry from Offshore Petroleum Development* Prepared for the East Coast Petroleum Operators Association IEPOA, St. John's, NL.

―――.1981c. "Fisheries Utilization in Eastern and Southern Newfoundland: An Assessment of the Impact of the Hibemia Development in the Period 1980-1990" *Report of the Fisheries Component - Environmental Impact Statement* Prepared for Mobil Oil Canada Ltd., St. John's NL.

―――.1983. *A Study of the Potential Socio-Economic Effects Upon the Nova Scotia Fishery from Offshore Petroleum Development* St. John's NL.

―――.1984 "The Newfoundland Fishery and an Assessment of Possible Impacts Associated with the Hibernia Development" *Background Report for the Hibernia EIS* St. John's NL.

Odum, H.T. 1971. *Environment, Power, and Society.* New York: Wiley-Interscience.

―――and E.C. 1981. *Energy Basis for Man and Nature.* New York: McGraw Hill. and Co.

O'Grady, Mary Anastasia. 2005 "Bolivarian Revolution = 'Cubanization' of Latin America", *The Wall Street Journal* [New York] Apr 29 p.A17.

Ohtsuki, H. C. Hauert, E. Lieberman and M. A. Nowak. 2006. "A simple rule for the evolution of cooperation on graphs and social networks", in *Nature* 441: 502-505 (25 May).

Okun, Arthur M. 1981. *Prices and Quantities: A macroeconomic analysis* Washington, DC: Brookings Institution.

Parenti, Michael 1996. *Dirty Truths: Reflections on Politics, Media, Ideology, Conspiracy, Ethnic Life and Class Power* San Francisco: City Lights.

Parkinson, Dave. 2005. "Road to 10,000 on TSX index greased with oil", *The Globe and Mail Report on Business* [Toronto] Jun 23 p.1.

Partridge, John. 2005a. "Fairbank wells in Ontario still pumping after all these years", *The Globe and Mail* [Toronto] May 21 pA1.

———. 2005b. "Sustained high prices redraw Canada's regional economic map", *The Globe and Mail Report on Business* [Toronto] May 21 p.18.

———. 2005c. "Loonie's rise greased as world is lured to new petro-currency", *The Globe and Mail Report on Business* [Toronto] Aug 12 p.1.

Patinkin, Don. 1956. *Money, Interest and Prices: An integration of monetary and value theory* New York: Harper and Row - 1965 edition.

———. 1969. "The Chicago Tradition, the Quantity Theory and Friedman", *Journal of Money, Credit and Banking*, Vol. 1, pp.46-70.

———. 1972. "Friedman on the Quantity Theory and Keynesian Economics", *Journal of Political Economy*, Vol. 80, pp. 883-905.

———. 1981. *Essays On and In the Chicago Tradition*. Durham, NC: Duke University Press.

Pearson, Karl W.1892. *The Grammar of Science*. London: Walter Scott.

Perkins, John. 2004. *Confessions of an Economic Hit-Man*. San Francisco: Berrett-Koehler. Petras, James and Henry Veltmeyer. 2001. *Globalization Unmasked: Imperialism ikn the 21st Century* Halifax [Canada]: Fernwood.

Petty, William. 1678. *Politicall Arithmetick* London.

———. 1662. *A Treatise of Taxes and Contributions* London.

Phillips, A.W. 1958. "The Relation between Unemployment and the Rate of Change of Money Wage Rates in the United Kingdom, 1861-1957", *Economica*, Vol. 25, pp. 283-99.

Pigou, A.C. 1933. *The Theory of Unemployment* London: Macmillan.

Pittman, Todd. 2005. "US turns an eye to oil-rich Gulf of Guinea", *The Globe and Mail Report on Business* [Toronto] Aug 08 p.5.

Pitts, Gordon.2006. "Taking A Stand- How One CEO Gained Respect", *The Globe And Mail Report On Business* [Toronto] Jan 31 p.8.

Pocock, J.G.A. 1957. *The Ancient Constitution and the Feudal Law: A Study of English Historical Thought in the Seventeenth Century.* Cambridge UK: Cambridge University Press.

Polya, George 1981. *Mathematical Discovery: On Understanding, Learning, and Teaching Problem Solving* Toronto: Wiley.

Pottinger, Matt. 2005a. "Cnooc's Fu Shows Complex Trends In Chinese Firms", *The Wall Street Journal* [New York] Jun 24 p.C4.

———. 2005b. "Aramco Discusses A Second Project Based in China", *The Wall Street Journal* [New York] Jul 12 p.A2.

Prior, D., M.M. Khan, M.R. Islam and F. Taheri. 2005. "A novel oil-spill cleanup barge", *Jordan International Chemical Engineering Conference V.* Amman, Jordan: JICEC'05, Sep 12-15.

Prowse, D.W. 1896. *A History of Newfoundland.* London: Eyre and Spottiswoode.

Quesnay, François. 1766. "Analyse de la formule arithmétique du Tableau Economique de la distribution des dépenses annuelles d'une Nation agricole", *Journal de l'agriculture, du commerce et des finances* Paris.

Radowitz, Bernd. 2006. "Brazil, Argentina and Venezuela Set $9.2 Million Plan For Study Of Gas Line", *The Wall Street Journal* [New York] Mar 13 p.A8.

Rahman, M.H, M.N.Wasiuddin and M.R. Islam. 2004. "Experimental and Numerical Modeling Studies of Arsenic Removal with Wood Ash from Aqueous Streams", Vol. 82: 968-977.

Rahman, S., 2006. Personal communication, Dalhousie University, Halifax, Mar 02.

Rampton, Sheldon and John Stauber. 2002. *Trust Us, We're Experts.* New York: Tarcher [Penguin USA].

———. 2003. *Weapons of Mass Deception.* New York: Tarcher [Penguin USA].

Ricardo, David.1817. *On the Principles of Political Economy and Taxation.* London: John Murray,.

Richer, Shawna. 2005. "Cape Bretoners look forward to return of smaller King Coal at Donkin Mine", *The Globe and Mail Report on Business* [Toronto] Jul 06 p.1.

Riesman, David. 1950. *The Lonely Crowd: A Study of the Changing American Character.* London and New Haven: Yale UP.

Robinson, Abraham.1966. *Nonstandard Analysis* Princeton UP.

Robbins, Carla Anne. 2006. "West Talks Tough With Iran, Treads Lightly", *The Wall Street Journal* [New York] Jan 23 p.A4.

Rogers, Raymond A. 1995. *The Oceans Are Emptying: Fish Wars and Sustainability* Montreal: Black Rose Books.

Rosie, G. 1975. *The Scramble for Oil* Edinburgh: Canongate.

Rostow, W.W. 1960. *The Stages of Economic Growth: A Non-Communist Manifesto.* Cambridge UP.

Samor, Geraldo. 2006a. "Brazil's Petrobras- Self-Reliant Or Pliant", *The Wall Street Journal* [New York] Apr 21 p.A7.

——. 2006b. "Brazil's Petrobras Halts Investment Planned In Bolivia", *The Wall Street Journal* [New York] May 04 p.A4.

Sampson, Anthony. 1976. *The Seven Sisters: The Great Oil Companies and the World They Shaped* New York:Viking/Bantam.

——. 1978. *The Arms Bazaar: The Companies, The Dealers, the Bribes: from Vickers to Lockheed* Great Britain: Hodder and Stoughton.

——. 1981. *The Money Lenders: Bankers in a Dangerous World* Great Britain: Hodder and Stoughton.

Samuelson, Paul A. 1970. "Maximum Principles in Analytical Economics – The Nobel Prize Lecture" 11 December 1970., in Assar Lindbeck, ed., *Nobel Lectures in Economic Sciences 1969-1980* Stockholm: Stockholm University, 1981.

—— and Robert M. Solow. 1960. "Analytical Aspects of Anti-Inflation Policy", *American Economic Review*, Vol. 50 2., pp.177-94.

Sayers, Michael and Albert E. Kahn. 1947. *The Great Conspiracy.* New York: Boni and Gaer.

Scheer, Robert. 1982. *With Enough Shovels: Reagan, Bush and Nuclear War* New York: Random House.

Schierow, Linda-Jo.2001. *The Role of Risk Analysis and Risk Management in Environmental Protection* (Congressional Research Service - Resources, Science, and Industry Division, Sep 6 - B94036).

Schumpeter, Joseph. 1939. *Business Cycles: A Theoretical, Historical and Statistical Analysis of the Capitalist Process* New York: McGraw-Hill.

Schwartzkopff, Frances and Benoît Faucon. 2005. "Kerr-McGee to Sell North Sea Assets", *The Wall Street Journal* [New York] Aug 09 p.C4.

Scoffield, Heather, Gordon Pitts and Greg Keenan. 2006. "Manufacturing Change: At The Crossroads, Adapt Or Perish", *The Globe And Mail Report On Business* [Toronto] Apr 27 p.8.

Shahed, Kalam 2002. *Ethnic Movements and Hegemony in South Asia* Bangladesh: Hakkani Publishers.

Shapiro, Rhoda. 2006. Personal communication about e-mail correspondence with Bradmans editor Michael Keating.

Shockley, William J. 1956. "Transistor technology evokes new physics – Nobel Lecture 11 December", in in *Nobel Lectures, Physics 1942-1962.* Amsterdam: Elsevier [1964].

Silver, Sara, Shawn Young and Leila Abboud. 2006. "Alcatel's Merger With Lucent Stirs Culture Questions; Paris-Based Giant Will Get American CEO, but French May Exert Greater Influence", *The Wall Street Journal* [New York] Apr 03 p.A1.

Simons, Geoff. 1994. *Iraq: From Sumer to Saddam* (London: St. Martins Press).

Simpson, Glenn R and David Crawford. 2006a. "US Investigates Critical Supplier Of Russian Gas", *The Wall Street Journal* [New York] Apr 2 p.A6.

———. 2006b. "Proliferation☐Of 'Shell' Companies☐Arouses Scrutiny", *The Wall Street Journal* Apr 25 p.A4.

Smith, Adam. 1776. *An Inquiry into the Nature and Causes of the Wealth of Nations* (Edinburgh).

Smith, Graeme. 2006. "Ottawa To Push For Gas Deal Between Petrocan, Gazprom", *The Globe and Mail Report on Business* [Toronto] Feb 13 p.1.

Smith, Rebecca. 2005. "California Takes Steps To Prevent Utility Shutoffs", *The Wall Street Journal* [New York] Oct 28 p.C4.

Solomon, Jay and Neil King Jr. 2005. "Iran Pipeline Complicates South Asia Policy", *The Wall Street Journal* [New York] Jun 24 p.A4.

Solow, Robert M. 1978. "Summary and Evalution", in *After the Phillips Curve: Persistence of high inflation and high unemployment* Boston: Federal Reserve.

Seed, T. 2006. "A Media of Progress, Enlightenment and Freedom is Possible", in Seed, T. et al. *Media and Disinformation – The Last Ten Years*. Halifax: New Media [in press].

Spencer, Herbert. 1857. "Progess: Its Law and Causes", *The Westminster Review*, Vol 67 (April), pp 445-447, 451, 454-456, 464-65.

Sraffa, P. 1960. *Production of Commodities by Means of Commodities*. Cambridge: Cambridge University Press.

Stackhouse, John. 2005 "The new energy shock", *The Globe and Mail Report on Business* [Toronto] May 21 p.3.

Stalin, Joseph V. 1952. *Economic Problems of Socialism in the USSR*. Moscow: International Publishers.

Stecklow, Steve. 2006. "Did A Group Financed By Exxon Prompt IRS To Audit Greenpeace", *The Wall Street Journal* [New York] Mar 21 p.A1.

Stevenson, James. 2006. "Hebron Partners Suspend Project As Talks With Newfoundland 'Stall'", *The Chronicle-Herald* [Halifax], Apr 04 p.E6.

Stiglitz, Joseph E., "Information and the Change in the Paradigm in Economics" [Prize Lecture 8 December 2001], in Frängsmyr, Tore, ed. 2002. *Les prix Nobel – The Nobel Prizes* Stockholm, Nobel Foundation. 472:540.

Stockdale, Scott 2003. *History's Greatest Fraud: German External Loan 1924* Light Years Communications Brantford Ontario ISBN 0-9732118-0-6.

Stonehouse, David. 2005. "Fears of supertankers plowing through the vista", *The Globe and Mail* [Toronto]Oct 11 p.A1.

Stueck, Wendy. 2005. "Soaring coal prices energise western Canadian output, shipments", *The Globe and Mail Report on Business* [Toronto] Apr 06 p.1.

Subrahmaniyan, Nesa. 2005. "China plans $3-billion refinery expansion", *The Globe and Mail Report on Business* [Toronto] Jun 20 p.8.

Sumerlin, Marc. 2006. "The Upside Of The Oil Curse", *The Wall Street Journal* [New York] Jan 10 p.A14.

Taber, Jane. 2005. "PM scolds Harper, Klein on US trade spat", *The Globe and Mail* [Toronto] Oct 13 p.A6.

Talley, Ian. 2006. "Oil Lubricates Oslo Bourse" *The Wall Street Journal* [New York] Jan 16 p.C8.

Tarbell, Ida M. 1904. *History of the Standard Oil Company* (New York: McClure Phillips and Co., 2 vols.).

Taylor, A.J.P. 1961. *The Origins of the Second World War* New York: Fawcett.

———. 1974. *Beaverbrook* Middlesex, England: Penguin Books,.

Taylor, Roger.2006. "Exact fate of Deep Panuke still a mystery", *The Chronicle-Herald* [Halifax] Jan 04 p.C1.

Texas Instruments Corp. 1974. *TTL Handbook*. Lubbock TX: Texas Instruments.

Thorndike, E. L. 1911. *Animal Intelligence*. New York: Macmillan.

Thucydides, 1954. *The Peloponnesian War* England:Penguin.

Tobin, James. 1963. "Commercial Banks as Creators of Money", in D. Carson, ed., *Banking and Monetary Studies* Homewood, Ill.: Irwin.

———. 1965. "The Monetary Interpretation of History", *American Economic Review*, Vol. 55 3., pp.645-84.

———. 1970a. "Money and Income: Post Hoc Ergo Propter Hoc?", *Quarterly Journal of Economics*, Vol. 84 2., pp. 301-17.

———. 1970b. "Rejoinder to Friedman", *Quarterly Journal of Economics*, Vol. 84, p.327.

———. 1972a. "Friedman's Theoretical Framework", *Journal of Political Economy*, Vol. 78 6., pp. 853-63.

———. 1972b. "Inflation and Unemployment", *American Economic Review*, Vol. 62, pp.1-18.

———. 1980. *Asset Accumulation and Economic Activity: Reflections on contemporary macroeconomic activity* Chicago: University of Chicago Press.

―――― . 1981. "The Monetarist Counter-Revolution Today: An appraisal", *Economic Journal*, Vol. 91 1. pp.29-42.

―――― . 1982. "Money and Finance in the Macroeconomic Process", *Journal of Money, Credit and Banking*, Vol. 14 2., pp. 171-204.

Traynor, Fiona and Tony Seed. 1999. "AIMS – A Fish Story…", *shunpiking 26* [Halifax, Canada].

Tuchman, Barbara W. 1962. *The Guns of August* New York: Random House.

Tuck, Simon. 2005. "Energy plan to include gas-price monitor", *The Globe and Mail* [Toronto] Sep 30 p.A1.

Turner, James S., ed. 1970. *The Nader Report: The Chemical Feast* New York, NY:Grossman.

Udall, Randy and Steve Andrews. 2001. *Methane Madness: A Natural Gas Primer* (Denver CO: Community Office for Resource Efficiency).

United Nations. 2004. *Statistical Review of World Energy*. New York: UNCTAD.

United States. 1999. "An overview and history of gas deregulation". Washington DC: Low-Income Home Energy Assistance Program [LIHEAP] Clearinghouse – Department of Health and Human Services.

―――― . 2004. *Annual Energy Review*. Washington DC: Department of Energy.

―――― . 2005. *Canada – Country Analysis Brief*. Washington DC: Department of Energy – Energy Information Administration [February].

Ushakov, Yuri V. 2006. "Don't Blame Russia", *The Wall Street Journal* [New York] Feb 13 p.A17.

Vallely, Paul . 2006. "Joseph Stiglitz: 'Politicians like Blair and Brown have given global poverty new prominence' - The Monday Interview: Former chief economist, World Bank", *The Independent* [London], Feb 20.

Vardanis, Christina. 2005. "A bitter wind blows into cottage country over wind-farm alt-energy scheme", *The Globe and Mail* [Toronto] May 21 p.M1.

Veblen, Thorstein J. 1909. "The Limitations of Marginal Utility", *Journal of Political Economy* Vol 17.

Veltmeyer, H and J. Petras. 2001. *Globalization Unmasked*. Halifax: Fernwood.

Vajda, S. 1981. *Linear Programming: Algorithms and Applications* London: Chapman and Hall.

Wadel, Cato. 1969. *Marginal Adaptations and Modernization in Newfoundland* St. John's: Newfoundland Institute of Social and Economic Research.

―――― . 1973. "Capitalization and Ownership: The Persistence of Fishermen-Ownership in the Norwegian Herring Fishery", in R. Andersen and C. Wadel edd., *North Atlantic Fishermen: Anthropological Essays on Modern Fishing* St. John's NL: Institute of Social and Economic Research, Memorial University of Newfoundland.

Wallace, David Foster. 2003. *Everything and More: A Compact History of* ∞ New York: Norton.

Wallerstein, Immanuel.1974. *The modern world-system: Capitalist agriculture and the origins of the European world-economy in the sixteenth century.* New York: Academic Press.

Walras, Leon. 1874. *Éléments d'économie politique pure, ou théorie de la richesse sociale* [Elements of Pure Economics, or the theory of social wealth] Lausanne.

Wang, Michael, Sarah Spikes and Simeon Kerr. 2005. "Middle East Oil-Cash Gusher Lures Foreign Investment Banks", *The Wall Street Journal* [New York] Oct 14 p.C6.

Warren, Susan and Jeffrey Ball. 2005. "A Change of Leadership For Big Oil Companies", *The Wall Street Journal* [New York] Aug 05 p.A3.

Watson, John B. 1913. "Psychology as the behaviorist views it", in *Psychological Review* [20]:158-177.

Webb, Sydney and Beatrice. 1920. *A Constitution for the Socialist Commonwealth of Britain.* London: Longmans and Green.

Website 1: United States Department of Energy – Energy Information Administration, http://tonto.eia.doe.gov/dnav/ng/ng_sum_lsum_dcu_nus_m.htm, last accessed 11 May 2006.

Website 2: Anadarko Petroleum, http://www.anadarko.com/, last accessed 11 May 2006.

Website 3: Grameen Bank, http://www.grameen-info.org/bank/GBGlance.htm, last accessed 13 May 2006.

Website 4: Institute of Near Eastern and African Studies discussion-blog, http://zennobia. blogspot.com/2006/01/last-jews-in-baghdad.html, last accessed 14 May 2006.

Website 5: Montreal Muslim News (transcript of television panel discussion) http://www.montrealmuslimnews.net/fulltranscript.htm, last accessed 14 May 2006.

Website 6: Petersen, Kim. 2005. "Disinformation: A Crime Against Humanity and a Crime Against Peace", *Shunpiking Online* Vol. 2 No. 6. Halifax [Canada]: New Media Publications, April. See http://www.shunpiking.com/ol0206/0206-mc-kp-miinfo-crime.htm, last accessed 22 May 2006.

Welton, Michael R. 2001. *Little Mosie from the Margaree: A Biography of Michael Moses Coady* Toronto: Thompson Educational Publishing.

Wessel, David, 2005. "Unlike Other Big Storms, This One Could Have Longer-Term Impact", *The Wall Street Journal* [New York] Sep 01 p.A1.

White, Gregory L. and Chip Cummins. 2005a. "Moscow's Plans In Energy Sector Rattle Foreigners; Ownership Limits Proposed For All Non-Russian Firms; BP Chief Lord Browne Meets Putin" *The Wall Street Journal* [New York] Apr 22 pA11.

———. 2005b. "Yukos Ex-Chief Sentenced to 9 Years", *The Wall Street Journal* [New York] Jun 01 p.A3.

———. 2005c. "Russia Turns Up Gas Pressure", *The Wall Street Journal* [New York] Dec 19 p.A15.

———. 2006a. "Flush With Oil, Kremlin Explores Biggest-Ever IPO", *The Wall Street Journal* [New York] Apr 18 p.A1.

Wicksteed, Philip H. 1910. *The Common Sense of Political Economy*. London: Macmillan.

Willis, Andrew and Patrick Brethour. 2006. "Oil Patch Expects Richer Shell HQ Buyout Offer", *The Globe And Mail Report On Business* [Toronto] Jan 03 p.1.

Wolfram, Stephen. 2002. *A New Kind of Science* Wolfram Media.

Wood, John H.; Gary R. Long and David F. Morehouse Energy Information Administration. 2003. "Long-Term World Oil Supply Scenarios – The Future Is Neither as Bleak or Rosy as Some Assert" United States. Department of Energy.

World Commission on Environment Development. 1987. *Our Common Future* ["The Brundtland Report"]. New York: United Nations, Doc. A/42/427.

Wysocki, Bernard Jr and Jacob M Schlesinger. 2005. "For US, China A Replay of Japan", *The Wall Street Journal* [New York] Jun 27 p.A2.

Yates, Frances A. 1978. *The Art of Memory* UK: Penguin Books.

Yergin, Daniel. 2006. "How Much Oil Is Really Down There", *The Wall Street Journal* [New York] Apr 27 p.A18.

York, Geoffrey. 2005a. "Demand China's unquenchable thirst", *The Globe and Mail Report on Business* [Toronto] May 21 p.19.

——— and Dave Ebner. 2005b. "Oil thirst from China adds fuel to trade tussle", *The Globe and Mail* [Toronto] Jan 14 p.A1.

———. 2005c. "Saskatchewan says China itching to acquire oil, uranium assets", *The Globe and Mail Report on Business* [Toronto] Jan 26 p.4.

———. 2006d. "Blowout In Bangladesh- Niko Resources' tale of woes", *The Globe and Mail Report on Business* [Toronto] Apr 01 p.1.

Young, Marilyn. 1968. *The Rhetoric of Empire: American China Policy, 1895-1901* (Cambridge, MA: Harvard UP).

Zatzman, G.M., I. Saney and M.R. Islam. 2003. "The Bush Doctrine: Hegemonism and 'Free-Market' Solutions", *Proceedings of the 6th International Conference of Economists and Accountants on Globalization, 9-13 Feb 2004*. Havana.

———. 1975. "American destroyer visit a 'first' since the end of WW2", *Evening Telegram* [St. John's NL] Aug 12.

INDEX

A

Aβ, 32
academic, 15, 18, 34, 35, 36, 54, 58, 65
academics, 54
acceptance, 18, 32, 54
access, 73
accidental, 14, 77
accountability, 51
accounting, 57, 74
accumulation, 50, 51, 52, 58, 62, 63, 65, 75, 76, 84
accuracy, 67
achievement, 6, 15, 16, 52
administration, 60
Afghanistan, 2
Africa, 59, 60
African-American, 35
age, 73, 79
agent, 74
agents, 51
agriculture, 69, 70
aid, 5, 15, 60, 61, 62
air, 25
alien, 16, 64
alternative, 25, 26, 27, 61, 68
alternatives, 26
Amazon, 25
amendments, 4
American colonies, 71, 72, 74
analysts, 42, 61
animals, 81
antagonist, 29
ants, 62
application, 11, 27, 63
applied mathematics, 8, 12
argument, 2, 14, 18, 27, 35, 62, 65, 74, 84
Arkansas, 81
armed forces, 60
articulation, 67
artificial, 23
Asia, 59, 60
assassination, 27
assault, 2, 36, 49
assaults, 1
association, 25
assumptions, 7, 13, 18, 42, 44, 54, 68
asymptotically, 64
Atlantic, 1, 37, 38, 39, 40, 41, 42, 47, 67, 72, 73, 78
atmosphere, 24
atomic physics, 61
attacks, 1, 2, 34
attention, 38, 84
attractors, 14
authority, 4, 6, 14, 17, 27, 35, 36, 52, 53, 80
automation, 66

B

backwardness, 59, 67
baggage, 64
baking, 75
barrier, 55
Bayesian, 33
Bayesian methods, 33
benefits, 35, 52
bias, 34, 55
Big Bang, 31
bilateral, 38
birth, 72
blame, 31
blaming, 48
boats, 72
body, 31, 38
Bolshevik Revolution, 62
boundary conditions, 7, 13, 21
Brazil, 25
Britain, 15, 26, 55, 58, 66
British, 11, 26, 67, 71, 72
bubble, 23
bureaucracy, 62
Bush administration, 25
business, 24, 45, 80, 83
bypass, 74

C

calculus, 4, 6, 8, 9, 13, 23, 50
Canada, 1, 25, 36, 37, 38, 39, 41, 43, 45, 49, 69, 72, 73, 78, 79
capacity, 37, 43, 84
capital, 2, 4, 16, 41, 47, 50, 51, 52, 58, 62, 64, 67, 68, 69, 70, 71, 72, 74, 78, 81, 82, 83, 84
capital accumulation, 58, 64
capitalism, 49, 50, 51, 54, 55, 58, 59, 63, 64, 66, 69, 71, 75, 78, 79, 80, 82, 83, 84
capitalist, 23, 48, 49, 50, 51, 58, 59, 60, 62, 64, 65, 66, 67, 69, 70, 75, 76, 77, 78, 79, 80, 82, 83, 84
capitalist system, 49, 51, 65, 69, 75, 77, 79, 82, 84

Carbon, 24
cartels, 66
Cartesian coordinates, 7
cast, 23
category a, 62
Catholic, 4, 6, 42, 52
Catholic Church, 4, 6
causal relationship, 32
causation, 36
celestial bodies, 4
cell, 75, 79
chaos, 21
chaotic, 14
Charles Darwin, 28, 29, 53
chemical, 24
chemistry, 30
chicken, 84
children, 35
China, 60, 69, 82
chopping, 63
circulation, 70, 71, 84
citizens, 61
civil war, 76
classes, 63, 72, 81
classical, 18
clean air, 25
climate change, 24
Clinton administration, 82
clothing, 74
Co, 29
CO_2, 24
coal, 73
coastal communities, 38, 41, 72
cognition, 1, 2
coherence, 54
Cold War, 2, 54, 62
colonial, 39, 43, 49, 58, 59, 60, 67, 69, 70, 71, 72, 80
colonial power, 58
colonial rule, 67
colonisation, 42, 59
combustion, 14
commerce, 42, 70, 71
commercial, 37, 38, 45, 47, 71, 72
commitment, 46, 48, 63

commodities, 71, 75, 76, 77, 78, 79, 84
commodity, 27, 28, 69, 70, 74, 75, 76, 77, 78, 79, 84
Communist Party, 58
communities, 38, 40, 41, 48, 69, 71, 72, 73
community, 62, 71
competition, 28, 34, 35, 54, 62, 67, 83
complementarity, 45
composite, 65
comprehension, 3
compulsion, 72, 77
computation, 20, 63
computer, 82
Computers, 82
computing, 10, 11, 65
conception, 16, 21, 29, 42, 59, 68
conduct, 44
conflict, 45, 83
confusion, 42, 47, 49, 60
Congress, iv, 54, 58
consciousness, 1, 2, 3, 26, 27, 38, 41, 47, 57, 66, 79
conspiracy, 84
construction, 59, 74
consumption, 61, 64, 77, 84
contempt, 62
continuing, 50, 61, 66
continuity, 5, 7, 10, 55, 56
control, 5, 36, 39, 42, 43, 70, 80, 81, 82
controlled, 35, 67, 74
convergence, 62
copyright, iv
corporate sector, 38, 73
corporations, 46, 48, 68, 72, 80, 81, 82
correlation, 31, 32, 33, 34
correlation coefficient, 32
correlations, 32
cosine, 9
coverage, 81
credit, 70
crime, 27
Cuba, 52, 65
cultural, 64
customers, 73, 82
cycles, 9, 14, 21, 50, 51, 52, 64

D

daily living, 17
death, 5, 53
debt, 58
deconstruction, 48, 49
deduction, 33
defects, 23, 75
definition, 5, 20, 63, 77, 79
degenerate, 33
degradation, 50, 74
degree, 6, 15, 41, 57, 65, 76
degrees of freedom, 56
demand, 28
democracy, 15
demographics, 34
denial, 1, 15
depression, 66
derivatives, 5, 6, 11
desire, 6, 81
desires, 1
detection, 41
developed countries, 25, 63
developing countries, 24, 25, 52, 54, 63
dietary, 74
differential equations, 6
differentiation, 5, 24
direct observation, 56
disaster, 37
discipline, 18, 39
discontinuity, 8, 19
discourse, 2, 47, 52
discreteness, 19
disinformation, 2, 14, 15, 24, 25, 31, 60, 75
displacement, 64, 67
distress, 73
distribution, 33, 84
division, 69, 75
domain, 5, 9, 21
dominance, 69
downsizing, 77
drying, 70
duration, 51
dynamic environment, 15, 55

E

earth, 15
ecological, 2, 37
economic, 1, 3, 14, 18, 19, 25, 27, 28, 29, 31, 37, 38, 39, 40, 41, 43, 45, 48, 50, 51, 55, 58, 59, 60, 61, 62, 63, 64, 66, 67, 68, 69, 74, 75, 78, 79, 80, 81, 84
economic activity, 41, 45
economic development, 50, 60, 61, 63
economic growth, 58, 66
economic policy, 82
economic resources, 28
economic systems, 37, 62, 68
economic theory, 1, 28, 38, 48, 63
economics, 16, 18, 26, 34, 35, 37, 40, 42, 43, 44, 45, 54, 57, 59, 64, 74
economies, 52, 59, 60, 62
economy, 23, 27, 28, 46, 59, 61, 62, 65, 67, 69, 74, 75, 78, 79, 83
ecosystem, 37
Eden, 59
education, 31, 34, 35
egg, 84
Einstein, 54
elaboration, 3, 7, 28, 30
electrical, 81
electronic, iv, 20, 81
electrostatic, iv
emergence, 18, 30, 36, 47, 49, 50, 53, 54, 57, 64, 82
emigration, 71
emission, 25
empowerment, 17
encapsulated, 5
encouragement, 58
energy, 3, 14, 17, 62
engineering, 8, 54
England, 4, 5, 69, 70, 71, 72, 74
English, 6, 36, 42, 45, 67, 71, 74
enslavement, 60
enterprise, 1, 3, 21, 32, 46, 68, 75, 78, 79, 84
enthusiasm, 66
entrapment, 59
entrepreneurial, 72

environment, 15, 18, 25, 43, 55
environmental, 2
environmental crisis, 2
equality, 59
equilibrium, 8, 13, 14, 15, 16, 18, 24, 55, 59
equilibrium state, 8
equipment, 64, 72
Euler, 6, 52
Eurocentric, 11, 49
Europe, 4, 14, 15, 54, 65, 69, 70
European, 1, 4, 16, 27, 28, 40, 41, 42, 47, 49, 50, 52, 54, 58, 59, 60, 64, 69, 70
Europeans, 40, 69
evidence, 24, 31, 32, 33, 35, 37, 41, 46, 48, 51, 57, 82, 83
evolution, 7, 18, 31, 53, 63, 67, 83
evolutionary, 18
evolutionary process, 18
exclusion, 52
exercise, 3, 62, 68, 80
exogenous, 69
expectation, 11
expert, iv
expertise, 48
experts, 48
exploitation, 40, 41, 66, 78, 84
exponential, 19, 52, 64
exposure, 31, 54
expression, 64
extraction, 21, 69, 84
extrapolation, 30

F

failure, 30, 39, 43
fairy tale, 84
faith, 32
false, 32, 33, 34, 43, 57, 62
family, 11, 80
famine, 71
fatherhood, 76
February, 58
federal government, 37
Feynman, 20
finance, 47, 81

fire, 46
First World, 54, 72
fish, 2, 37, 38, 41, 45, 48, 69, 70, 72, 73, 74
fisheries, 2, 36, 37, 40, 41, 43, 45, 46, 47, 50, 52, 69, 71, 72, 74
fishing, 2, 37, 38, 40, 43, 45, 48, 67, 68, 70, 71, 72, 73, 74, 79
focusing, 19
food, 2, 30, 45, 69, 72, 74
food production, 30
foodstuffs, 41
foreign aid, 60, 61
forestry, 69
forests, 24, 25
fossil, 53
framing, 31
France, 15, 58, 60
freedom, 15, 48, 53, 56, 72
freezing, 23
fresh water, 25
fruits, 77
frying, 46
fuel, 14

G

Galileo, 4, 14
gas, 75
General Agreement on Tariffs and Trade, 73
generation, 26, 31, 57, 66, 73
geography, 79
George Berkeley, 5
global warming, 24, 25
God, 5, 29, 30
goods and services, 28
Gore, 82
Gore, Al, 82
government, iv, 2, 15, 25, 37, 38, 40, 42, 48, 51, 55, 59, 61, 62, 73, 78, 82, 83
government intervention, 82
government policy, 78
graph, 8, 9, 29
gravitation, 28
gravity, 4
Great Britain, 15

Great Depression, 54
Greeks, 15
grouping, 28, 32, 60
groups, 43, 59, 80, 81, 82
growth, 30, 58, 63, 66
Guinea, 24

H

hands, 41, 74, 77
Harvard, 33, 34
harvest, 38
harvesting, 38, 41
head, 17, 52, 58, 69, 75
health, 31
heart, 35, 46, 49
hedging, 20
height, 2
hiring, 70
hole argument, 35
human, 2, 4, 5, 14, 15, 17, 19, 21, 24, 27, 29, 35, 37, 43, 45, 52, 63, 81
human agency, 2
human animal, 81
humans, 52
humility, 21, 56
Hungary, 58
hypothesis, 35, 46

I

idealism, 29, 68
ideas, 53
identification, 61
ideology, 49, 55
illusion, 20, 55
images, 19, 20
immigration, 73
immortal, 53
imperialism, 80, 83
inauguration, 60
incentives, 24
income, 41, 72
incomes, 38

independence, 20, 71, 72
independent variable, 50, 51
India, 69, 82
Indian, 11
indicators, 24, 52
individual development, 13
Indonesia, 58, 60
industrial, 15, 26, 28, 30, 49, 61, 64, 66, 69, 71, 75, 78, 81
industrial production, 71, 78
industrial revolution, 26
industry, 2, 26, 40, 42, 44, 60, 62, 69, 71, 73, 74, 78
inequality, 59, 62
inertia, 14
inferences, 43
infinite, 5, 11, 14, 84
influence, 4, 58, 80
infrastructure, 25
injunction, 20
injury, iv
injustice, 62
input, 35, 55
instinct, 24
institutions, 35, 36
instruction, 20
instruments, 46
intangible, 3, 16, 19, 21, 26, 27, 36, 42, 43, 47, 49, 52, 63, 69, 74, 76, 78
integrated unit, 47
integrity, 14, 20, 21, 32, 33, 36, 44, 45, 56
intelligence, 27
intensity, 32, 41
intentions, 27, 28, 42, 45, 47, 51, 57
interaction, 43
interest, 8, 17, 19, 21, 26, 27, 42, 47, 51, 57, 60, 63, 65, 72, 77, 78, 80, 81
interest groups, 80, 81
interference, 42
Intergovernmental Panel on Climate Change, 24
internal combustion, 14
international, 39, 41, 43, 63
Internet, 82
interpretation, 18, 37, 42

interrelations, 8
interrelationships, 37
interval, 8, 10, 42, 50
intervention, 27, 28, 58, 61, 62, 82, 83
interview, 23, 24, 25
invariants, 41
inventories, 73
investment, 21, 50, 52, 61, 69, 78
investors, 72
invisible hand, 26, 27, 28, 29, 42
Iraq, 15, 25, 59
Islam, iii, 20, 21, 33, 48, 53
island, 38
isolation, 20
Israel, 58

J

jobs, 25
judgment, 35, 47
judiciary, 36
jurisdictions, 82
justice, 62
justification, 18, 27, 46, 50, 61

K

Keynes, 18, 19
Keynesian, 35
kidnapping, 70
knowledge, 2, 6, 17, 32, 36, 52, 53, 57, 63

L

labour, 2, 5, 37, 40, 43, 45, 46, 66, 69, 72, 73, 74, 75, 76, 77, 78, 79
labour force, 40
land, 74, 80, 82
large-scale, 71
Latin America, 58, 59, 60, 61
laughing, 30
law, 14, 19, 23, 26, 28, 29, 30, 31, 66, 77
laws, 3, 4, 6, 19, 23, 28, 29, 30, 46, 47, 68, 83
laws of motion, 3, 4, 19, 28, 46, 68, 83

Index

layoffs, 77
lead, 36, 37, 78, 79
leadership, 58
Lebanon, 59
Leibniz, 5, 6
liberal, 60
lifespan, 21
lifetime, 11, 55, 57
likelihood, 33
limitations, 3
linear, 4, 6, 7, 9, 20, 21, 51, 67
linear model, 20
links, 39
listening, 17
living conditions, 41
location, 34, 68, 80, 81
logical reasoning, 30
London, 23, 59
long run, 18
long-distance, 82
long-term, 1, 17, 27, 28, 45, 52, 57
lying, 9, 32, 39, 49

M

machinery, 83
Madison, 34
magnetic, iv
mainstream, 2
maintenance, 14
management, 80
manipulation, 49
mantle, 44
manufacturing, 72
marches, 25
marginal utility, 67
marginalisation, 52
marginalization, 38
market, 24, 25, 34, 35, 41, 55, 69, 70, 71, 72, 82
markets, 41, 61, 67, 72, 73
Marx, 15, 25, 26, 30, 44, 47, 49, 53, 54, 65, 66, 69, 70, 71, 76, 83, 84
Marxism, 44, 49
Marxist, 17, 44

mask, 49
mass, 38, 50, 61, 77, 79
materialism, 68
Mathematica, 1, 5, 28
mathematical, 4, 5, 11, 18, 19, 20, 21, 29, 55, 56
mathematicians, 5, 6, 11
mathematics, 4, 5, 8, 9, 11, 19, 63
meanings, 42
measures, 19, 54
meat, 35
mechanical, iv, 8
mechanics, 19, 30, 54
membership, 23, 51
men, 6, 47
mentor, 11
mergers, 82
metaphor, 27
methodology, 19, 35, 36
metric, 45, 51, 65
militant, 11
military, 1, 58, 60, 67
Milton Friedman, 35
mining, 36, 73
misappropriation, 19
misleading, 43, 56
misunderstanding, 30, 75
mode, 3, 14, 30, 45, 64, 69, 70, 75, 76, 77, 78, 81, 83
modeling, 10, 11, 20, 53, 56, 58
models, 13, 18, 19, 20, 23, 52, 55
modern capitalism, 51
modern society, 79
momentum, 6
money, 24, 28, 71, 73, 77, 78
monopoly, 1, 54, 67, 68, 74, 80, 81, 82, 83, 84
moratorium, 37, 73, 74
mortgage, 72
Moscow, 39, 54
motion, 3, 4, 6, 14, 15, 17, 19, 20, 28, 29, 44, 46, 59, 68, 70, 78, 83
movement, 1, 48, 50, 79, 80
multiplicity, 47, 56

N

narratives, 28
nation, 34
national, 34, 39, 72, 79
NATO, 60
natural, 6, 8, 9, 10, 15, 18, 19, 20, 23, 26, 28, 29, 30, 31, 35, 40, 43, 53, 54, 55, 56, 57, 60, 62
natural environment, 18, 43
natural laws, 6, 29, 30
natural resources, 60
natural science, 8, 15, 18, 26, 30, 35, 54
natural sciences, 8, 18, 26, 35, 54
natural selection, 53
needs, 17, 30, 52, 60, 80, 84
negative consequences, 31, 47
network, 15
New England, 69, 71, 72
New World, 42, 43, 49, 69, 70
New York, iii, iv, 35
New Zealand, 35
Newton, v, 1, 3, 4, 5, 6, 7, 8, 9, 14, 20, 23, 28, 29, 52, 54, 68
Newtonian, 13, 21, 28, 29, 50, 51
Nobel Prize, 35
Non-Aligned Movement, 58, 60
non-linear, 7, 9, 18, 20, 21, 53, 56, 67
non-linearities, 9
non-linearity, 18
non-profit, 61
non-uniform, 50
normal, 13, 27, 66, 68
norms, 32
North America, 40, 60, 71, 72, 79
North Atlantic, 50, 69
NS, 74
nuts, 3

O

objectivity, 46
observations, 10, 46
obsolete, 80
offshore, 38, 69
oil, 75
old-fashioned, 45
opacity, 33
operator, 6, 7, 9
opinion polls, 15
opposition, 16, 81
oppression, 74
orbit, 14
organization, 78
output, 9, 40, 41, 55
overpopulation, 30
overproduction, 50, 77, 84
ownership, 41, 54, 62, 76, 80, 81

P

paper, 34
Papua New Guinea, 24
paradox, 15, 50
parents, 34, 35
passive, 52
patents, 81
pathways, 24, 55
peers, 18
pendulum, 20, 60
perception, 4, 29
performance, 34
periodic, 9, 21, 30
permit, 9, 21
personal, 32, 36, 80
personality, 58
perspective, 42, 44, 83
petroleum, 24
philosophical, 4, 28, 68
philosophy, 11, 16, 45
physical sciences, 13
physics, 30, 61
piracy, 70
planetary, 6, 24
planets, 4
planning, 48, 62
plants, 72, 73
play, 27, 44, 51
policy-makers, 27

political, 1, 14, 23, 34, 37, 39, 45, 46, 49, 51, 54, 55, 58, 59, 60, 64, 65, 80, 83
politics, 15, 26, 35, 37, 39, 40, 42, 45, 58, 59, 62
polynomial, 9
pools, 73
poor, 23, 31, 34, 61
poor performance, 34
population, 30, 31, 34, 40, 51, 71
population growth, 30
positivist, 16
posture, 6
poverty, 62, 63
power, 27, 43, 54, 58, 61, 64, 76, 77, 80, 81, 82, 83
powers, 48, 59, 60
pragmatic, 9, 81, 82
pragmatism, 2, 68
prediction, 53, 66
preference, 46
prejudice, 14, 18, 32
preparation, iv
preparedness, 25
presidency, 25
president, 60, 81
pressure, 17, 19
prices, 73
primacy, 55
principle, 1, 23, 27, 28, 38, 45, 51, 57, 61, 84
pristine, 64
private, 6, 43, 53, 54, 58, 61, 62, 65, 75, 76, 77, 83
private investment, 61
private ownership, 54, 62
private property, 53, 65, 77
private sector, 61
private-sector, 61
probability, 19, 32, 35, 54
probability distribution, 19, 32
procedures, 32, 35
producers, 47, 69, 70, 71, 76, 79
production, 20, 27, 30, 41, 47, 51, 52, 54, 59, 62, 63, 64, 67, 69, 70, 71, 75, 76, 77, 78, 79, 81, 83, 84
production function, 47

productivity, 41
profit, 1, 16, 40, 66, 78, 84
profits, 48, 64, 67, 78
progressive, 48, 72
promote, 26
property, iv, 15, 36, 53, 62, 80
proposition, 82
pseudo, 14
psychology, 7
public, 6, 15, 26, 32, 33, 34, 35, 43, 53, 58, 62
public interest, 26
public opinion, 58
public schools, 33, 34, 35
public transit, 32

Q

qualifications, 25
qualitative differences, 79
quantum, 19, 30, 31, 53, 54, 64
quantum mechanics, 19, 30, 54

R

radical, 44, 49
rain, 25
rain forest, 25
random, 29, 34, 68
range, 5, 14, 26, 36, 60, 63
rationalisation, 77
raw material, 2, 45, 60, 64, 73
raw materials, 45, 60, 64
RBOCs, 81
real time, 43
reality, 6, 9, 13, 14, 15, 20, 21, 23, 31, 32, 39, 44, 46, 60, 81, 83
reasoning, 30, 32, 33, 47, 51, 62, 66
recessions, 66
recognition, 56
reconcile, 66
reconciliation, 84
reconstruction, 43, 48
recruiting, 80
recurrence, 51

reduction, 54, 65, 67
reference frame, 54
refineries, 75
refining, 24
regenerate, 84
regional, 2, 37, 48, 59, 61, 73, 75, 81
regular, 60
rejection, 45
relationship, 23, 26, 32, 38
relationships, 23, 26, 45, 52, 71, 76
relativity, 54
religious, 5, 23, 32, 35, 53, 75, 79
Renaissance, 4, 61
rent, 78
replacement, 54, 66
reproduction, 75, 77
reputation, 35, 54
resale, 77
research, 1, 13, 19, 30, 34, 42, 43, 44, 53, 57
research design, 34
researchers, 4, 6, 11, 13, 14, 16, 35, 37, 40, 55
resistance, 1, 14
resolution, 57
resources, 28, 35, 52, 60, 74
responsibility, 31, 48
revenue, 26, 74
revolt, 4
revolutionary, 54, 62, 63, 64, 66, 67
rewards, 24
rhetoric, 64
rights, 72
robbery, 69
Royal Society, 4
rural, 30, 70
rural population, 30
Russia, 25, 54, 60
Russian, 62, 73, 74

S

Saddam Hussein, 15, 25
salary, 15
salt, 69, 72
satisfaction, 66
scalar, 65

schema, 6, 20
scholarship, 17
school, 20, 33, 34, 35, 75, 79
science, 3, 4, 6, 13, 14, 15, 16, 19, 23, 24, 29,
    30, 31, 32, 34, 35, 36, 42, 43, 44, 45, 50,
    52, 53, 54, 55, 65, 68, 78
scientific, 1, 3, 4, 14, 15, 19, 20, 21, 28, 29,
    30, 31, 32, 33, 36, 43, 44, 45, 46, 50, 52,
    54, 78
scientific method, 29, 32, 44, 50, 52
scientific theory, 31
scientists, 5, 19, 28, 30, 34, 37, 49
scores, 34
search, 71
secret, 27, 28, 52
secrets, 20
security, 26
seeds, 16
selecting, 13, 31
self, 14, 26, 46, 52, 73
self-interest, 26, 52
sensitivity, 21
sentences, 76
separation, 15, 42, 45
series, 10, 53, 80, 82
service provider, 82
services, iv, 27, 28, 61, 82
servitude, 72
shares, 59, 70
shortage, 2
short-term, 17, 18, 19, 24, 28, 40, 45, 57
shoulders, 41
sign, 18
signs, 67
similarity, 64
sine, 8
slavery, 74, 79
slaves, 69, 70, 74
Slovenia, 51
social, 1, 2, 13, 15, 16, 17, 18, 19, 26, 27, 30,
    31, 32, 34, 35, 36, 37, 38, 39, 40, 41, 42,
    43, 44, 46, 48, 50, 51, 52, 53, 54, 55, 57,
    59, 61, 62, 63, 64, 65, 67, 68, 71, 72, 75,
    76, 77, 78, 79, 80, 81, 83, 84
social capital, 51

Index

social class, 63, 72, 81
social development, 16, 81
social events, 19
social justice, 62
social life, 62
social movements, 59
social order, 15, 31, 53
social phenomena, 16
social relations, 67, 78, 79, 81
social responsibility, 52
social sciences, 18, 19, 26, 32, 35, 36, 53, 54, 57
social services, 61
social transition, 51
socialisation, 52
socialism, 55, 58, 59, 63
socialist, 55, 58, 59, 62, 65, 67
socially, 28, 52, 75, 77, 84
society, 16, 23, 26, 27, 30, 35, 50, 51, 52, 63, 65, 76, 78, 79, 80
sociologist, 16
sociologists, 37
sociology, 37
solutions, 11, 13, 20, 21, 24, 34, 56
sorting, 48
sounds, 75
sovereignty, 38, 72
Soviet Union, 2, 15, 38, 58, 59, 60, 62, 73
Spain, 42
specialists, 31
speciation, 31, 53
species, 18, 31, 38, 53
speed, 4, 14
speed of light, 4
stability, 21
stages, 28, 59
standard of living, 41
stasis, 55
State Department, 15
state intervention, 27, 62
state planning, 62
statistics, 19
steady state, 9, 13, 14, 15, 20, 46, 55
stock, 30, 73, 80
streams, 33, 34

strength, 71
subjective, 35, 42
subsidies, 74
subsidy, 74
subsistence, 71
substitution, 81
Sun, 68
superiority, 15, 26
superposition, 6, 7, 53
supplements, 68
supply, 14, 28, 30, 61, 73, 74
suppression, 54
surplus, 16, 38, 64, 66, 67, 71, 72, 73, 76, 77, 79, 84
survival, 26
symbolic, 63
symmetry, 11
sympathetic, 54
syndrome, 1, 26, 27, 31, 44, 46, 48, 49
Syria, 59
systematic, 29, 77, 78
systems, 20, 37, 47, 59, 62, 68, 75

T

takeover, 77
talent, 62
tangible, 3, 16, 21, 26, 27, 63, 68, 74, 76, 78
task force, 47
teaching, 34, 36
technology, 20
telecommunication, 82
telecommunications, 82
telephone, 81
television, 81
television coverage, 81
temporal, 4, 9, 18, 19, 30, 31, 36, 43, 44, 47, 49, 50, 51, 52, 54, 64, 65
territorial, 38, 39
test scores, 34
theoretical, 5, 7, 18, 23, 24, 27, 54, 58, 63, 68, 75, 78
theory, 1, 5, 6, 7, 18, 24, 26, 28, 31, 35, 38, 48, 49, 53, 54, 57, 58, 59, 61, 63, 64, 66, 68, 69, 73, 75, 79

132 Index

thinking, 17, 55, 57
Third World, 24
threat, 24
threatened, 69
timber, 79
time, 1, 2, 3, 4, 5, 6, 7, 9, 10, 11, 13, 14, 15, 16, 17, 18, 19, 20, 21, 25, 27, 29, 30, 31, 32, 34, 35, 36, 37, 38, 40, 41, 42, 43, 44, 45, 46, 48, 49, 50, 51, 52, 53, 54, 55, 56, 58, 60, 62, 64, 65, 66, 68, 69, 72, 73, 78, 79, 81, 83
Tony Blair, 25, 55
toxic, 24
trade, 25, 26, 72, 79
trading, 24, 25, 69
tradition, 23, 79
traffic, 43, 69
training, 66
transactions, 75, 78
transformation, 41, 43, 62, 63, 67, 70
transformations, 43
transition, 72
transitions, 51
trawlers, 72
trawling, 73
trend, 16, 36, 43
trust, 15, 81, 82
trusts, 81
turbulence, 19
two-dimensional, 11
typology, 63

U

uncertainty, 19, 54
undifferentiated, 79
unemployment, 77
unfolded, 46
uniform, 50, 54
uniformity, 21
unions, 35
United States, 15, 36, 59, 60, 62, 66, 71, 72
universality, 68
universe, 29, 46, 54
universities, 54

untainted, 11, 45
Uruguay, 73
Uruguay Round, 73

V

values, 5, 8, 9, 50, 52, 77
variable, 50, 51, 76
variables, 3
vector, 6
velocity, 6
vessels, 74
vice-president, 82
Vietnam, 60
vision, 63, 64, 74
vouchers, 34, 35

W

wages, 64, 72
walking, 25
Wall Street Journal, 33, 35
war, 25, 72, 76, 80
Warsaw, 58
Warsaw Pact, 58
Washington, 39
waste, 67
water, 25, 33, 60
watershed, 3
wealth, 1, 37, 43, 45, 50, 52, 62, 72, 76, 77, 80
wear, 64
West Indies, 69
wheat, 79
wilderness, 74
winning, 46
wireless, 82
Wisconsin, 34
wisdom, 48
wood, 60
words, 23, 29, 47, 58, 68, 70, 81
work, 1, 2, 3, 4, 5, 7, 9, 14, 15, 16, 18, 19, 20, 30, 32, 34, 37, 38, 41, 43, 44, 47, 49, 53, 54, 56, 58, 66, 67, 68, 71, 78, 84
workers, 19, 41, 61, 70, 77, 84

workforce, 40, 73
working conditions, 41
World War, 54, 67, 72
World War I, 67
World War II, 67
writing, 36

yield, 5